An Astronomer's Tale

A Bricklayer's Guide to the Galaxy

Gary Fildes

arrow books

1 3 5 7 9 10 8 6 4 2

Arrow Books
20 Vauxhall Bridge Road
London SW1V 2SA

Arrow Books is part of the Penguin Random House group of companies
whose addresses can be found at global.penguinrandomhouse.com.

Penguin
Random House
UK

First published by Century in 2016
First publishing in paperback by Arrow Books in 2017

www.penguin.co.uk

A CIP catalogue record for this book is available from the British Library.

ISBN 9781784754389

Printed and bound by Clays Ltd, St Ives Plc

Penguin Random House is committed to a sustainable future
for our business, our readers and our planet. This book is made
from Forest Stewardship Council® certified paper.

Dedicated to the working people of the North-East: the builders and the plumbers, the nurses and the soldiers, the mams and the dads. If you once dreamed of space, of being able to fly to the moon, dare to dream again.

Contents

Author's Note

As well as chronicling my life story in astronomy, this book offers seasonal bi-monthly stargazing guides to the constellations, which include star charts. Structured into alternate chapters between the autobiographical narrative, these bi-monthly night-sky chapters can be read in or out of sequence, and each one spans four constellations that can be observed in the northern hemisphere, as well as one other celestial object.

Below is a key to these bi-monthly night-sky sections. Elsewhere in the book, in the autobiographical narrative, I will also discuss other celestial objects and practical tips about astronomy.

For additional information, at the back of the book is an 'Annotated Glossary' and a brief 'Note on Equipment'. To get the most out of the night-sky sections, I recommend that newcomers to astronomy turn to the glossary before they begin the book, as it explains concepts such as 'magnitude' and describes some of the terminology that I use in the book. There is also a 'List of Constellations' and a short section titled 'Where Can I Get More Help?', which suggests other learning resources.

KEY

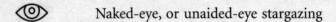 Naked-eye, or unaided-eye stargazing

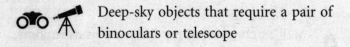 Deep-sky objects that require a pair of binoculars or telescope

Preface

It is 5.45 p.m. and I get in my car and leave home as usual. I always enjoy this hour-long drive away from the urban sprawl of Newcastle, away from the houses and the busyness of the city. A calming silence descends on me as the car's headlights shine out in front, guiding me along the empty country lanes. I join the A68, or the 'Military Road', a reminder of times long gone: it was originally laid by the Roman army and used as a passage between York and Scotland. I trace the winding route as it cuts through Hadrian's Wall, dissecting vast cornfields, and continues up and over ancient hills. Stretching out far in front, the hedge-lined road leads me unerringly north. I open the driver's window slightly and feel the air. It's fresh and cold and fast. As I bound over the crest of another hill, the front wheels briefly clear the ground. On the descent I get the first glimpse of open countryside.

It is bursting with life. A pair of deer pauses fleetingly; pheasants totter by dangerously close to the roadside. The fields to my left are populated with tiny low-lying clouds of white wool; the lambs are appearing. A few spring and trot alongside the hedges until they spot the car and race off to Mam and the safety of her back. Soon pine trees surround me. The Sitka spruces are tall and grand and line the road like sentinels. They signify something special for me. Dark-Sky Guardians, as I like to think of them. I have been transported to the plains of Scandinavia. Gently rounding another corner, I see water like glass off to my right. The lake is vast. I marvel that it is man-made and built in the heart of Kielder Forest in Northumberland. Apparently it is the largest in Europe. An osprey soars high above and in the distance I can just make out what appears to be a flock of birds flying off towards the approaching night. It is already getting dark as I strain my eyes through the windscreen to see the first stars appearing from the twilight. Only the hum of the car's engine reminds me that I have work to do, that I have a group of visitors waiting.

It's just past 6.45 p.m. when I slam the door and walk along the sandy gravel path. I'm buoyed up by the latest weather forecast, suggesting it will be a clear evening. The air is crisp and I take another big lungful as I habitually raise my head skywards. Ursa Major, or the Great Bear, gently growls at me, shining brightest. Radiating out from within the constellation is an asterism in the familiar shape of a cooking utensil. Her body and tail constitute the 'pan and its handle', seven stars that resemble a dot-to-dot of what you would find on your kitchen cooker. I can see the bright blue hue of Alcor and Mizar, two distant energetic balls of hot rarefied gas that form a part of the handle of the 'pan', held together by gravity. My eye then finds two stars in the head of the pan, Merak and Dubhe. They are important: they point the way to Polaris, the North Star, the position in the sky that represents our planet's rotational axis. As you might guess, it also points directly north, so if you get lost at night, knowing its position could save your life. I watch as Merak and Dubhe twinkle gentle hues of yellow and amber and then blue again as our atmosphere distorts the faint light from these distant suns. I know the constellation that contains this asterism, or pattern of stars, well; it's familiar and I feel welcomed, as if receiving a respectful nod from across the room by an old friend. Even if that friend is a bear.

The sky is changing again now as darkness falls faster and the Sun sets over the horizon. I can see the bright glare from Venus as it reflects the day's remaining sunlight through its thick, cloudy atmosphere of carbon dioxide. It is so bright an array of colours dances in the golden light of the sunset. Above is Jupiter, shining softly and not as brightly. At over 350 million miles away, this behemoth of a planet, the largest in our solar system, follows Venus effortlessly down towards the horizon. But it is now nearly 7 p.m.; I really have to get to work.

As I approach the observatory, the inky black sky stretching out around it, I hear in the background the comforting whirl of the wind turbine which powers our observatory and array of telescopes. The timber structure is perfectly silhouetted against Deadwater Fell in the east with a distant listening post for the RAF perched on its summit. The two square observatory turrets that house the telescopes rise proudly from the base of the building, separated by the observation deck used to stare up into the cosmos. In some ways, the observatory resembles a futuristic ship sailing out into the universe, with its minimalist shapes and clean corners dominating the surroundings. The larch cladding, which protects the telescopes residing in the turrets, is standing and fixed vertically, nailed into a defensive poise.

The guests are already there, sitting down and snuggling up on the two-person Moon chairs out on the deck under the stars. I can hear their excited chatter. A faint glow of red light is visible, shining up from our 'dark-sky'-sensitive light bulbs. This strange scarlet light preserves the visitor's now dark-adapted vision. The volunteers are animated. I can see Austin pointing at the sky, looking up eagerly. His head and neck are pushed to the extreme as he strains to see distant objects. Forget health and safety, I get a little gush of happiness as I realise that creaking necks are ignored and regarded as part of the job.

Tonight we are holding an event aimed at hunting out deep-sky objects like galaxies and star clusters that lie at huge cosmological distances from Earth. People flock to the observatory for many reasons, perhaps to see something rare or unexpected, such as the Northern Lights, or aurora borealis, which regularly dance in reds and greens in our northerly skies. Others come to get a glimpse of something bigger, to feel their own insignificance. Set in the vastness of space, guests often say, they feel small – but to me we are all significant. Tonight we hope to see everything, observing through our telescopes vast metropolises of stars and clouds of glowing gas. What we cannot see we debate and discuss.

At approximately 7.45 p.m. I welcome the forty guests inside the observatory. The small rectangular room we call the classroom can only hold forty people at a squeeze and tonight we are, as usual, at maximum capacity. In the corner the log-burning stove is keeping us warm and glowing amber. The indoor red dark-sky lights create an atmosphere of secrecy, as if we are all children hiding beneath a duvet cover. On the screen I'm showing pictures in high resolution of astronomical objects, many taken by the Hubble Space Telescope, others taken here at Kielder. I give a brief welcome and cover the usual ground rules, as well as pointing out the way to our composting loo. I talk about what we hope to see in the coming hours, especially the distant galaxies – those huge, often seemingly swirling patterns of billions of stars that lie so far away that the light they emit takes millions of years to travel to us. I talk in particular of a galaxy called the Whirlpool that lies around 22 million light years away. For effect I emphasise that the light from the Whirlpool galaxy has to cover 186,282 miles every second (the speed of light) for 22 million years, before we can even see it.

I lead the group outside to stand on the observation deck. I ask if anyone can see a faint fuzzy patch in the sky, due north. One guest spots it and then another. It is the Andromeda galaxy, a distance of 2.2 million light years away. It's the most distant object in the northern skies that any human being can observe without a telescope. I pick out with my laser pen some of the constellations that adorn our surroundings. In Greek mythology, Perseus was the first hero, the son of Zeus who became immortal after his death. Like the myth, Perseus the constellation harbours a favourite object, the variable star Algol, which pulsates over time. Climbing higher into the sky, I point out the Milky Way, our home galaxy. It steals the show, distinguished with knots of obscuring dust lanes meandering through its star fields.

There are a few gasps; our view looks quite different from the perspective of the sky you see from most cities. Due to light pollution in urban areas, on most clear nights you can see a maximum of thirty stars. But nestled within the Kielder Forest, away from any nearby metropolis, the observatory sits under the darkest skies in Britain. It is the third largest area of protected dark skies in the world, and with great darkness comes great clarity: thousands upon thousands of stars can be seen. They seem to reach out into eternity, as if a divine artist had painted them.

I explain to the group that there are many hypotheses about the naming of our galaxy, the Milky Way. The ancient Ionians theorised that it was milk spewing from the breast of Queen Cassiopeia, a beauty whose vanity led to her downfall. Others believe in stories of fireflies and godly artistry. Scientifically, of course, the naming of our galaxy doesn't matter, but we associate so much meaning with a name and it fuels our questions. What is the Milky Way? Why do we see it the way we do? If it is overhead in the sky, how can we say that we, on Earth, are inside it? What about the stars we can see to the left and to the right which clearly aren't inside the Milky Way, or are they? It all seems very confusing.

I say to the group, which ranges from children to pensioners – all united in their curiosity – that I like to think of galaxies as cities of stars. In the same way that people live in cities, stars live in galaxies. However, there is no city on our planet that possesses a similar number of people to the number of stars in a galaxy. Anywhere between 100 billion and 300 billion stars reside in the flat spiral disc we call our Milky Way galaxy. And of course each 'star' is a sun like our own.

I ask everyone to imagine that we have a big circular

disc like a Frisbee lying in front of us on the ground. On it we should paint the stars and swirling patterns. This is what a telescope does: observes a distant galaxy from the outside. Next, I ask all the guests to change things around and get inside the pretend galaxy. We can start by cutting out the centre and standing in the middle of it, so it surrounds our feet. Then, by lifting up the galaxy, like a hula hoop, it gets higher and higher, closer and closer to our eyes. We still see it as a spiral, until the one magic moment when we lift the pretend galaxy up to eye level. What do we see? No longer is it a spiral circular shape at all, but a thin ring of Frisbee. Now we are in the galaxy. It is the same when we look up from the observatory and see precisely what any person who has ever looked up at the Milky Way witnesses: our galaxy appears as a thin band across the sky. Why? Because we are inside it.

Follow-up questions about the celestial Frisbee rightly come thick and fast from the group – it has taken me decades to get my own head around the concept. But just as I start to explain again, I notice a few fingers take to the air in a different direction. A lady standing close by me suddenly shrieks. 'Oh! Look at that aircraft coming towards us.' I turn my head in the direction of the voice, and I can see it now too. I don't have to strain into the darkness; I can see what appear to be bright landing lights. The group instinctively huddles together on the observation deck, guarded either side by the two turrets. Some people are nervous, but I still hear the sound of Moon chairs scraping across the deck as some visitors try to get the best vantage point possible. It is 9.30 p.m. now and the sky is as dark as it will get, at least until the Moon sets later. Thousands of stars shine down on us, and even several satellites that can be seen with the naked eye are spinning across the sky. But we are transfixed by what appears to be an aircraft. Seconds later it seems to be moving slowly or hovering, we can't quite tell. More seconds elapse and the group becomes ever more attentive. This is no aircraft. As the object seemingly hurtles towards us, its colour brightens. The gasps and cheers grow louder; it is by now easily the brightest thing in the sky, even outshining the Moon.

'Oh my God! Is it going to hit us?' The unidentified flying object silently fizzes amber to green, so bright and colourful, then blue, then amber again. By this time fifteen seconds have passed since the initial observation. People are snatching at their bags to get at cameras. I fall silent. It began in the south-east, and now it is due south. Looking at it square on, we can see it has developed a tail trailing behind it. It, too, burns and fizzles more colours, as forces begin to burn its surface to a crisp. In a flash it is now in the south-west and we can see debris falling from the tail. 'Keep watching!' I yell. It fades gradually, as gas and more debris fall like amber fairies diving down towards the Earth. We are all amazed, and then it is gone.

It lasted for twenty-five seconds from first sighting. Research later suggested that the object, a bolide meteor the size of a bus, had entered the Earth's atmosphere over Belgium and progressed through, burning up over the Atlantic Ocean. Observers from Europe and Ireland also reported sightings of this extremely bright meteor before it disappeared.

Asteroids like this can be leftovers from the formation of our solar system some 4.5 billion years ago. Normally they are held in a stable gravitational orbit around our home star, the Sun, tucked safely in a region called the asteroid belt in between the orbits of Jupiter and Mars. However, a collision in space or a little nudge in this region can change that stable orbit and send these objects hurtling towards us. These objects can be travelling between 11 and 70km per second and weigh thousands of tonnes. With such huge energy of motion, or kinetic energy, they pose huge risks to us on Earth.

This one, however, wasn't dangerous. But neither was it predicted or logged or tracked – we knew nothing of its arrival. Once it interacted with our dense atmosphere, its fate was sealed. Frictional forces converted its kinetic energy to heat and light, and we saw the process unravel before our eyes. At such great temperatures, sometimes

3,000–4,000 degrees, the object is changed forever. Some pieces may land on Earth and be found as meteorites; most are burnt away by our protective atmosphere. What we saw was an everyday occurrence on Earth since the dawn of time. The way our planet is peppered by rocks from space may even have seeded the building blocks for life on Earth in the distant past. But for tonight we are content to look and marvel at the spectacle. No Ph.D.s in physics are needed, nor an expensive education: we simply look and enjoy Mother Nature reminding us who is boss.

I walk back into the classroom and to the log-burning stove, where a group of children are enthusiastically describing what they have seen. Before we resume the night's stargazing, a thought flashes through my mind. For twenty-five years I was a bricklayer working on building sites, lost but searching for something. It feels good to be home at last.

INTRODUCTION

How to Build an Observatory

'Areet, Terry.'

'Ow, Gary.'

'Have I told you about the obsy?'

'Eh?'

'I want to build an observatory, but I'm skint.'

'Oh . . . right.'

'See those bricks over there, the ones lying on the ground?'

'Yeah.'

'Can I have them?'

It is April 2000 and another day at work on the building site. Terry laughs out loud. He must think I'm a bit brazen. The bricks in question belong to our employer Bellway Homes, a developer operating across the north-east of England. Terry is an estimator, part of the technical team that costs up the price of making a building. I am a bricklayer. We get on well, particularly as he is also an amateur stargazer himself, so I know I am playing to a home crowd.

I explain that the planned observatory is the dream-child of the Sunderland Astronomical Society, or the SAS as we call ourselves. Our motto is 'Who Stares Wins'. I've been a member for three years, meeting every Sunday in one of the upper rooms we rent in the large Victorian terraced house on the seafront owned by the Quakers. I was first invited along by my next-door neighbour Dickie, who was already part of the group. One weekend, he discovered from my son that I was hiding a huge passion for the stars.

'Dad, Dad!' my son Graham shouted excitedly as he ran into our house. 'The bloke whose car I've just washed has an LX200, the same one you've been looking at in the magazines!'

'What?' I pulled back the curtain to look in the direc-tion of Dickie's house. 'He has an LX200? Are you sure?'

I was flabbergasted as my son ran back to tell Dickie that his dad had a telescope too. That was it. Dickie and I became fast friends and he welcomed me into the fold. I immediately knew the group were my kind of people. The chair, Don Simpson, would usually lead the evening's proceedings. Don was tall and thin and could usually be seen modelling his USS *Nimitz* baseball cap, tight jeans and a pair of storm-trooper black boots. He would sit at the front of the room with the committee members either side of him and roll a fag. With the filter in his mouth, staring at his rolling handiwork, he would utter, 'I change me arn oil 'n drink me beer outta cans; there's nee public school ties in here.' We were a tight-knit crowd of fifteen or so: nerdy, diverse and utterly engrossed in astronomy. From planning group trips or talks for the general public to sharing advice about telescopes or photography, our gatherings were the highlight of my week. I would look forward to them whilst working on the building site during the day, and dream of a clear night in the forests so that we could go off together with our scopes. Because I liked him so much, and he lived opposite me, I would pester Dickie relentlessly, until he later admitted that, whenever there was a clear night, if the phone rang, he would say to Anita, his wife, 'It'll be Gary.' But secretly I knew he loved it and he couldn't wait to get out there either.

For the society, this would be our very first observatory. Our plan was to build something small but permanent, which we could use to deliver stargazing experiences to the general public. The building itself would only house a single telescope through the roof, and only three to four people at a time could fit inside. But it would certainly be a little drier and warmer than the observation nights I had been giving to the public for the last year outside Kielder Castle, which could be sub-zero. The new observatory would be open to the whole community, inviting people of all ages to come and try the stars for themselves.

Its size was partly dictated by the dimensions of a white fibreglass observatory dome that had been donated. Three metres in diameter, it resembled a builder's hat, and once we had scrubbed off a layer of green algae, it was fit for purpose. Now it just needed a brick structure to sit upon, and a concrete base. Thanks to my job, I was uniquely qualified to volunteer my services.

Someone once told me that if you don't build your own dream, someone else will probably hire you to build theirs. For over two decades I had very literally been doing the latter, but I was proud of it. Life as a bricklayer was exciting at first. Mastering the trowel wasn't easy, and it took a lot of practice to mimic the more experienced hands. You needed to get the right amount of mortar, or 'muck' as it was known, on your trowel, before spreading it just thickly enough so that when you placed your brick down a little muck would squeeze out of the bed and you could scrape it off. If you did this correctly, the brick would be airtight and would bond with the muck. There was an immense feeling of satisfaction in the repetition, and an even greater sense of well-being many bricks later when I had finally learned the craft. The building site itself was also an enjoyable environment, full of my kind of people, the same lads I had grown up with on the council estate in Grindon village. Most of us were young and fit and strong, and we could take the physical punishment. I got to know Shaun Stokoe, a friend of my older brother, particularly well. He was one of the hardest workers and the best hod carrier I've ever known; he had the key job of carrying bricks around the site. One hod, or hoddie, would normally support two of us brickies as we built a wall. The hod Shaun used was a three-sided metal box attached to a long shaft. It could be filled with around fourteen bricks, which could weigh close to 40 kg when full. He would hold the hod shaft with one hand as he loaded up with bricks using the other, before slinging the box on his shoulder and lugging it to where I or another brickie was working. Shaun is still doing this job to this day.

The two of us were a well-matched team and soon we were scouting around sites to see who was paying the best wages. The summers were the best. In the late 1980s and early 1990s it wasn't unusual for us to be working onsite with a group of other lads, everyone in just a pair of Speedos and a set of steel-toed boots. What a vision. Bare-backed and bare-chested we would toil all day in the sun, returning home to our family sunburnt and worked out. Winters were not as pleasant. One year we were doing contract work for Barratt Homes. We were self-employed at the time, which was better for the money, but if you didn't work you didn't get paid – even if it was freezing cold. With four kids now to provide for, and nearing Christmas, the pressure to earn was intense. I bundled up warmly every day to fight the biting winds, but that didn't solve the problem of the ice-cold bricks. Bricks have a knack of getting wet, and can absorb enough water to nearly double in weight. As they're rough to the touch, the tips of my fingers were already skinless after laying hundreds of them per day. One morning I went to pick one up. It was hard and cold and so I grabbed it more vigorously than usual. When it didn't budge, I hit it with my hammer and up it came. But there it stayed; I couldn't get it off. The water had frozen my lukewarm fingers to the brick's surface. I shook my hand, which only made matters worse as it was too heavy for that. Eventually I tore my fingers away with my other hand, and off came the skin; or rather it remained attached to the brick. In exquisite pain, I screamed up into the grey hanging sky, heart pounding, and I remember thinking, get used to it, you are here to stay.

I was back on the site again the next day, and the day after. Without brickies and this army of workers who endure these conditions all year long, we would have no homes to live in. I knew this and felt proud to be doing the work. But internally I felt numb and the dream of astronomy from my youth was a distant memory. What I was doing now I would be doing until I could do it no longer. As I grew older, things got tougher each year. Backache is probably the worst part of the job. After a shift I would need to lie down on the floor at home and the kids would stand in the small of my back, to ease the pain and stretch me out. Afterwards I would gingerly lower myself into a hot bath.

I had tried everything to get out: selling insurance,

selling fish, even making applications to the police. But I could never find an escape. Although the winters could be hard, few jobs around Sunderland could match the money. As well as the physical toll, the thought preyed on my mind that somehow this life wasn't for me. I had internalised my passion for astronomy since I was a child; around most of the lads at work I was Gaz the lad, Gaz the Sunderland football fan. Most had no idea that by day I was trowelling, and by night I was reading physics books. The new observatory finally presented me with an opportunity. How strange that one craft was steering me towards another. Towards a life that had been my passion all along.

Back on the Bellway Homes building site, Terry was sold on my request for free materials, partly because of the good PR I promised it would reflect back on the company. And it worked – they generously donated the lot, from bricks and mortar to concrete. The foundations were eventually dug out and the concrete arrived. Reinforcement was put in and we filled the trench with the concrete. I will never forget the sight of the new observatory that first morning, with the tubs of ready-mixed mortar standing close by, a blue tarpaulin covering the other essential materials. Yes, it was a familiar sight to me – it looked like a building site – but this one was different, it really mattered to me. Work was about to start as I stood excitedly with a can of yellow spray paint and my tape measure, preparing to mark out the building line. Building walls isn't easy, especially when it is a radius wall, as we say in the trade, meaning it's a curved full circle. To achieve this you need precision, and your tools: a good trowel from which you can evenly spread your mortar, and a very accurate spirit level. 'Setting out', as it's called, is probably the most integral part of the process because once the 'line' is set, there is no going back on the dimensions. I started off by measuring from the centre of the floor to my outside wall. We had decided that the ring on which the 3-metre dome would sit and rotate would be positioned on the inner wall; this way we could maximise space inside the observatory. The wall would be 100mm deep, the width of the cavity would be 75mm, and the width of the outer leaf of bricks was also 100mm. Some simple maths later and we had discerned that the radius measurement, which was half of the floor, was 1.775 metres to the edge of the outer wall. I kept measuring from that central point and had soon spray-painted my circular base.

The bricks we were using were predominantly russet brown, smooth to the touch and each one around 225mm long, as old bricks were in the UK. However, there was our first problem. To construct a perfect circle of our size with rigid lumps of clay as hard as iron and 225mm long would not work. We needed the wall to curve smoothly, and with the bricks being the size they were, their long lengths would be too large for the radius. A better option would be to use smaller bricks, but how when we only had the pile we had begged? The solution was to cut them – every single one of them straight in half. For the next few weeks a very understanding friend, Mick McLaughlin, also a brickie, set about helping me snap each brick. We were still doing our normal jobs, so we would usually arrive in the evenings from our other sites. Of course, snapping a brick isn't like snapping a biscuit. It's more like hitting it precisely with a hammer. A brick hammer is a special tool and not like the everyday hammer you would use to knock in a nail. It is around the same size but is very strong and made of cast steel with a thick plastic handle. The first thing you realise when you use it is that it is very well balanced. It has to be, so that when you swing, it connects well and all of the weight and momentum is transferred into breaking the brick. It has one flat end and then tapers to an edge, which directs the force onto the brick. If you strike it cleanly, then you can expect a snapped brick in one go; sometimes it takes two attempts. A loud crack and a jolt up your arm usually signifies a successful hit. But like most bricklaying it is a skill that takes time to perfect; it had taken me at least a year or two to master the brick hammer when I was an apprentice. Thankfully, the muscle memory was still there, and a few weeks later each observatory brick was separated into two perfect little halves. We began to lay the wall and the smooth circle quickly took shape. Now we could play with the design.

I wanted to represent some of the major constellations in the pattern of the bricks. Fortunately we had been given a small assortment of yellow bricks, so once we had cut these too, we decided that each bright star in our design was to be represented by a yellow half brick, standing out from the russet-brown 'space' of the others. The first and most important constellation to design was Cygnus. Cygnus the Swan is a favourite among stargazers in the northern hemisphere. Sitting inside the plane of the Milky Way, it is found in the brightest region of our galaxy and is relatively easy to recognise due to the positions of its most brilliant stars, which together form the crossed shape of the royal bird of its namesake (its shape also gives rise to its other nickname, the Northern Cross). Cygnus' most brilliant star of all is Deneb, a blue-white supergiant 200,000 times more luminous than our own Sun, but over 2,000 light years from Earth. Deneb, located in the left-hand side of the constellation, forms the tail of the swan in flight. The star that forms the head is called Albireo or the 'beak star'. This appears to be one star but is in fact two: a small telescope would resolve two distinct points of light, one of which is vivid orange and the other bright blue. Track back from Albireo to Deneb and you will encounter the star Gamma Cygni, which marks the centre of the swan's back. From there, you can trace left and right out to the tips of the swan's wings. When I observe Cygnus I always think of a white swan flying through the star fields of the Milky Way.

We had decided to name our building the Cygnus Observatory because it was to be located in the grounds of the Washington Wetlands Centre, on the outskirts of Sunderland. Although located in an urban environment, not a dark-sky site, the bird sanctuary was still a good location for our first observatory, situated as it is in a relatively serene area away from lamp posts or other housing lights. Its wooden buildings are topped with moss-covered roofs and surrounded by gently sloping banks and grassed areas. A small lake marks the centre of the grounds, and birds fly overhead continually; Canada geese add a splash of colour, competing with the population of pink flamingos. The centre is perhaps most well known for its swans, though, so our observatory would always be called 'Cygnus' and the brickwork seemed a suitable gesture. We had yellow bricks spare so we could design Orion too, as well as other constellations.

Over the remaining summer months the observatory took shape. I was so proud to see the dome fitted, affirming the building's presence on the landscape. We painted its fibreglass shell and it shone brilliant white. As six months on the project came around, and the warm summer evenings came to a close, the final part of the build was of course the most important: the installation of the telescope. The scope that had been donated by Durham University was a 14-inch Newtonian Cassegrain focus. It is a popular design, and the name comes from the famous British physicist Sir Isaac Newton and an engineer who revolutionised the optical system. Laurent Cassegrain, who was a French Catholic priest, published the first notable design for a reflecting telescope in 1672.

We agreed to put our precious new telescope on a concrete pier that would stand in the centre of the observatory. The day it was installed, we were tired but overjoyed. With this powerful tool we could comfortably observe the planets and bright deep-sky objects. A few weeks later, on 2 October 2002, the Cygnus Observatory was officially opened to the public. The ribbon was cut by Professor Sir Arnold Wolfendale, a noted British astronomer who served as the Astronomer Royal in the early 1990s, and who is now Emeritus Professor in the Department of Physics at Durham University (Sir Arnold would go on to become a friend and a mentor, eventually the man to bestow upon me an Honorary Master's degree in Astrophysics from Durham years later). But it wasn't the decorum of the opening that was the greatest source of my pleasure. It was that something so special had come from a place of such passion among my friends, and from such humble means.

*

Now, nearly fifteen years later, I have been fortunate to help build another observatory – the Kielder Observatory.

Cygnus got me started in astronomy and taught me the art of the possible. Kielder has transformed my life.

When you mention the word 'observatory', what springs to mind? A clinical building perhaps, with a white hemispherical dome. Inside, it is probably cold and dark and full of complicated equipment. Only the glow of a computer screen illuminates the hypnotised face of the pale creature peering into the telescope. An astronomer. A geek.

I like to think of observatories and astronomy in general in different terms. To me, an observatory is a little home-away-from-home for us all to play in. It protects us from the elements and keeps us warm, all the while revealing the distant wonders of the universe. They are places to be enjoyed by all, regardless of your age or level of education.

Like our mission at the Kielder Observatory, I hope this book can in some way whet the appetite of readers to explore the night sky for themselves. It is the story of my own journey of becoming an astronomer. The stars have been a constant thread in my life since child-hood, but there have been times when I lost direction. My earliest memory related to astronomy is Christmas Day 1970 when I was five years old. That morning I remember crawling under my dad's tree and looking up through the branches. The magic started instantly, the smells triggering Santa Claus, ginger wine, tangerines and marzipan. The tree, however, was what I liked the most. It was enchanting, full of colourful twinkling lights. I found myself lying on my back and wriggling beneath the tree to change my perspective or just to get closer, to get in. I was transported to a different world. I would see a faint blue glow from a bulb, a red starburst and then a green one too. Sometimes the lights seemed distant and harder to see, so I had to move my head to look beyond the twisting pine needles and branches. I became locked in a trance that I never wanted to come out of. My own universe.

That was where it started. The study of light informs all astronomy. My brother Anthony got a telescope a few years later for Christmas which he didn't use, but I did. I saw this big thing in the sky – I didn't know it was the Moon at first. My interest in astronomy flourished as I grew up, but I kept quiet about it. I love Sunderland – the place, the football team – and I'm massively proud of where I come from. But it wasn't a particularly nourishing environment to grow up in being interested in science. My parents couldn't have been more loving or caring, but like me they didn't know what to do about it and I wasn't encouraged at school either. So I internalised my passion.

At that time in Sunderland you either worked down the pit, the shipyards or on a building site when you left school. I ended up in one of them. I married young, had four beautiful children and got to grips with my 'trade'. In many ways I was happy and I moved forward with my life. But there were other periods when I loathed the cycle I was stuck in. For years I kept my feelings about science bottled up like a guilty secret. By day working on a building, while at night inhabiting a twilight world with a few like-minded friends. I got to my mid-thirties and thought, 'I've got to change this.' I came out. As an astronomer.

In this book I describe how I changed career mid-life, how I taught myself and how I learned from others to discover one of life's most fulfilling pleasures. I hope if there are any readers out there who find themselves in a rut, regardless of lifestyle or career, they might find some common ground. I also want to share my passion. Getting away from our desks and our smartphones and out into nature to spend time under the sky is an endlessly fascinating, and to my mind, rewarding, way of life that I think many people would enjoy if they tried it themselves. I know there is an appetite. The original remit for Kielder was to host four to six events a year and fill the classroom, which holds forty people – so 240 visitors at most annually. By the end of the first year we had had 1,200 visitors. This year we will have received 24,000 stargazers. If – or hopefully when – we expand with a new extension, we expect visitor figures to top 75,000 a year.

I have been given many incredible opportunities to travel around the world with astronomy to see great skies and meet many extraordinary people, from reclusive observers in Chile to astronauts in Houston. I am extremely grateful to all of the people who have helped and encouraged me on my way, and I have chronicled my experiences in an attempt to convey the rich variety of life that this pursuit can offer. Finally, the book includes a month-by-month calendar guide to the constellations. Readers can learn how they can measure distance to these dots of light, and how study of these stars relates to more daunting topics such as space-time and the physical size of the universe. Astronomy can seem daunting, and I know some people, myself included for a very long time, can get defensive, even fearful that they don't have permission to learn. My feeling is that we can all understand some of how the universe works, but as I often say to the visitors of our observatory, if all you want to do is look at the sky, then that's good enough, that's OK.

Join me now as I go back to the very beginning of my journey in astronomy. It was the first step on a long and wonderful road, but like so many passions it was hardly an auspicious start.

1

First Light

First light is a cherished term in astronomy. It
denotes the special moment when a telescope and
its observer embark together on a maiden voyage. Both
pointed skywards, light enters the instrument and hits
its optics for the first time. A virgin image of the heavens
is reflected into the stargazer's eyes.

For the old hand, or for those few who may have
even built a telescope, this initial peer into the eyepiece,
into the unknown, is accompanied by much nervous
excitement. Will it work? I bloody hope so. For an
observatory it can be the crowning moment of years of
hard work. Although first light is rarely spectacular – like
most things in life, a scope needs constant calibration
to obtain the best results – one normally trains a new
telescope on the brightest object in the sky first. At the
opening of the Kielder Observatory in 2008, this object
was to be Vega, found in the constellation Lyra. This

bluish-white star, which is the fifth brightest in the night sky, was climbing high in the north-east that April. The blue photons emitted from Vega's core didn't disappoint, making for a dazzling christening, the star's hydrogen particles fusing into helium twenty-five light years away from Earth. But whether you are a trained astronomer using an expensive piece of kit, or trying your brother's new toy for the first time, as I did when I was nine years old, first light is invariably a seminal moment.

My initiation was on Christmas Day, 1974. On my way upstairs to bed, finally out of our parents' clutches and ready to tuck into our bounty of chocolates and sweets, I caught a glimpse of it. The light was off in the bedroom, but I could see the shape of my brother Anthony's biggest present, silhouetted against the tall window. Long and slender, with three legs, it pointed out of the window, which had ice and frost on the inner pane. Despite the cold and the eerie light cast from outside, I was tempted into the semi-darkness. I put a hand on the telescope and found the source of light pooling in: a giant Moon. Being nine, I knew Neil Armstrong had walked on it when I was little, and rockets had flown there. Looking down at Anth's new toy, I knew the tube and the Moon were somehow connected. I wondered if this strange machine could be the spaceship to take me there.

At that point my younger brother Marty ran into the room shouting and trying to play, continually flicking the light on and off. Despite not having the faintest inkling of how to use the thing, I sensed the lights weren't helping. I shoved Marty out and picked up the telescope from its tripod. I peered through the bottom of the tube. All I could see was a blur. So that's what this thing did: look at a blur. I liked it immediately.

Anth's telescope, as I know now, was a refracting instrument with a very low-quality lens at the front of the tube. I was staring through the bottom of the tube without an eyepiece, so I would never have been able to see any stars, or much else for that matter. Nothing that the telescope was designed to see. But I was happily unaware, transfixed by what must have been out-of-focus dirt and dust. This detritus was being magnified many times into beautiful patterns and hues of grey and blue, scattered into my eyes. I kept moving the telescope, still trained on the sky, for what must have been about an hour, until, finally, a light so bright it seemed astonishing to me flashed past my line of sight. Staring again into the telescope, going cross-eyed, I got a glimpse of the unfocused Moon. It was fleeting but very different from what I had previously been looking up at with my naked eyes. I had to find it again, so I swept left to right then back again, until bingo, I found the brightness once more.

'Marty, Marty! Wake up. Look at this. Look. It's the Moon!'

My mind was alive and firing with ideas and questions;

I knew there must be more to this blur machine than I had first thought. Over the coming hours and days I discovered and fitted the eyepiece, mostly by chance, and slowly learned how to control the basic movement of the scope. The Moon became an elusive friend. I would often find it, only for it to disappear ten minutes later, without my even touching the telescope. It would be years before I discovered the real reason behind this: that the Moon's position across the sky alters because of the Earth spinning on its axis. As the Earth spins, the mount for the telescope, myself and the house where I lived were also moving. I didn't know any of this yet, but I knew I had caught the astronomy bug.

I wanted to share my ground-breaking discovery of the Moon with the world, or at least with my family and friends. It wasn't too much of a stretch for my dad, who was an out-and-out Trekkie and inclined to science fiction. A mechanic by trade, he would love nothing more than to come home and watch an episode of James T. Kirk wooing his next alien into submission. My dad would record each episode with his small cassette tape recorder. Positioning the microphone in front of the TV, he would order my brothers and me into silence for the full thirty-five minutes. Later, when he played the audio recordings back in the car, we would snigger when we heard one of us say 'ere man, shuddup' or 'ow man, gerrof is'. Dad would look up from the wheel and smile ever so slightly. Since my revelation about the

Moon, he was pleased that my latest hobby might keep me out of trouble. I always think how grateful I am now as an adult that my dad bought that telescope; he didn't have to, but something in him also connected to astronomy and he must have wanted that for us too. I wish I could ask him.

Growing up in Grindon, a council estate on the outskirts of Sunderland, I knew nothing of the class divide as a kid, or of any other way of living. As far as I was concerned, our house was the only one in the universe that mattered, the only home in our neighbourhood that always had Christmases, holidays, brothers and both a mam and a dad. I had no boundaries and could play outside with my friends whenever I wanted, spending most nights playing football, knocky-nine doors with my gang, or having scraps with the 'Thornies', the boys from Thorny Close who were our sworn enemies for no other reason than they lived in another part of town. We would meet on the sandhills and have it out: furtive punches and kicks, lobbed bottles and insults. It was thrilling as my gang fought together as one, running bravely into battle, then scurrying away if we heard a siren. As long as you were going with the tide, you rarely got hurt.

My passion for astronomy was growing and I mentioned my telescope to my best mate Paul Lundy, or Lun as I called him. He was baffled and characteristically asked if we could play 'internash' instead, a jumpers-for-goalposts variation of football. Not put off, later that evening I was sitting on Mrs Thomson's wall with the rest of our gang, waiting with a 2p piece in my hand for Tognarelli's ice-cream van to come by. The Moon had just broken out from behind the clouds and I pointed it out excitedly to the rest of the gang, some a few years older than me who I hadn't seen before. I tried to show them the craters that gave the Man in the Moon his familiar puffed-out expression, but I never got to finish ... laughter broke out all around. Instinctively I knew the laughter could change quickly, so I went silent. I can't remember exactly what they said, but next there was a lot of swearing, which became tugging at my coat. I knew what was coming. I remained sat on the wall, staring at the ground as I watched the shoes circle in around me. Lun, I sensed, was moving away. Who could blame him? A punch in my ear and then a kick was all it took. I was on the ground, curled up in a ball.

I found out later they were from another part of town called Pennywell, where the really rough boys lived. After they had left, my friends picked me up, frantic and crying, and took me home. It was becoming apparent that I had better keep the Moon to myself if I didn't want to get into trouble. So long as I fought the Thornies, played football and swore a lot with the gang, I was fine. Well, it wasn't fine for me, but lessons were being learned. I was being conditioned to live my life a certain way, even if I wasn't fully aware of it.

School was an interesting environment for me. There was control and structure, whereas away from school I played out all night and the rules I followed were those of the gang. Of course I pushed against these boundaries, which meant I didn't have a great relationship with many of my teachers. I was mostly interested in being the focus of attention, spending my time cultivating a 'Jack the lad' persona in order to be accepted by my peers. For my efforts I was rewarded with plenty of appointments with the bamboo cane in the head teacher's office, and detention in the library, but it was here I discovered that school could be a fount of information. I read about physics, the one lesson I loved. Something about it felt useful, and of course we occasionally veered into the territory of astronomy. Because I rarely did the required reading for class, my teacher got angry when I asked obvious questions, shutting my enthusiasm down in its tracks. Partly for this reason, and the less than stellar response from my friends in Grindon, I kept my passion for the stars to myself. I was regarded as a classic case of an intelligent child who lacked application – 'Gary could do better'. But so could my teachers, I thought. On the whole they seemed to lack enthusiasm. We were all set for our lives; we were factory fodder and they knew it. Why should they bother? If I was going to learn anything useful, I thought I would have to do it for myself.

At school I mostly continued in the same vein, trying

at some things and not being bothered with others. There were no interventions by any teachers that would help me move in a certain direction. Life was a waiting game – waiting for something to happen – but I wasn't empowered to think that I could control or influence my life in any real way; I was treading water and blamed no one for that. All of our careers information was focused on careers in the military or in shipyards. I don't want to sound like some sort of bleeding heart who was helpless and had no influence over my life; on the contrary, I was strong-minded and I thought I knew best. I knew no other way than to be strong, to get on with it. My family unit remained sturdy and it was undoubtedly the one place where I felt completely safe and achieved the most pleasure. Family provided the only model I was motivated to re-create. I was in a typical phase for a teenager: bored and uninterested. The only temptations were money or buying a car, or a new girlfriend; to do any of that I needed a job.

I left school at sixteen in 1982 with a smattering of qualifications. Without a thought about sixth form or university, I wandered straight to the dole office. I never thought of further education. Why would I? It just didn't seem worthwhile. If I moved into a blue-collar job, then I would belong to the streets and the gang and my mates; this way I would be accepted and not singled out. I really do remember having those thoughts; my astronomy was a secret to keep quiet about. After all, my existence wasn't a bad one, my mates were amazing and I loved them all; and my family was everything to me. However, I wanted something else that back then I just wasn't aware of. I didn't hold anyone responsible, it just worked out differently for me and the academic path wasn't a reality. My future was Sunderland and I would do Sunderland things whether I liked them or not. More times than not, I liked them back then.

My memories of stargazing during those years were on the street corner waiting for the ice-cream man on Mrs Thomson's wall. I would sit looking at the sky, dreaming about distant stars. I lived a lot in my head and wondered continually about the universe and space; it was not an escape from my family but more of a desire to explore. I don't think I knew then how to vocalise what was quite a complicated set of feelings.

When I practised astronomy on my own, I recall seeing the stars as a jumbled-up mess. Pretty, yes, but randomly scattered across the sky. One night in particular sticks in my mind. We were playing football in a back lane against some garage doors and the dim street lights were acting as floodlights. I knew it was a clear night and I wasn't much of a footballer. Tognarelli's ice-cream van was in the area, but from where we were playing we couldn't see it, so I offered to go and 'keep watch'; that way I could perch on Mrs Thomson's wall and stare as usual at the sky. The stars were flickering particularly brightly and I remember being struck by the range of whitey blues and yellows and oranges; I was hooked. Normally I was overwhelmed by the sheer number of dots of light and had no idea where to look, or how to jump from one star to the next without losing my place when I blinked. But that evening I started to focus on a single constellation I had read about, and then on a nearby star to locate it.

Things began to make sense as I navigated from one point to another; for the first time I was seeing the sky as an organised structure, starting with Ursa Major and Polaris. Over the next few years I would hone this technique, taking my astronomy to the next level from these two locations in the sky. I'll share my approach now with you over the rest of this chapter, but it's worth explaining some of the underlying science first, as well as the myths, which can be distracting. My core knowledge as I was growing up was patchy; if I had spent more time reading books I'm sure it would have rapidly speeded up my stargazing.

*

Step out on any starry night in the northern hemisphere and one group of stars is always visible: Ursa Major, or the Great Bear. It is one of the most recognisable constellations in the sky, the third largest, and most commonly known for its seven stars in the shape of a giant saucepan, known as the Plough or the Big Dipper. It was this characteristic shape I would stare at from Mrs Thomson's wall.

As it appeared to me, and for many observers in the northern hemisphere Ursa Major is circumpolar – it never sets below the horizon and so is visible all year round. It rotates around the northern celestial pole: in spring the handle points down and in the summer it points west. Spring, however, is a particularly good time for observing it as it sits high in the northern sky and is most easily recognisable. Like all constellations, it was first mapped thousands of years ago when ancient civilisations across the world drew pictures in the stars. As a teenager I could never really see these shapes and I would often get confused thinking about the Greeks. But I do remember understanding why they were important for our ancestors, particularly farmers and land workers. They had recognised that constellations were cyclical, the same year after year, so by identifying which constellations were rising and setting, they could plan to plant and harvest crops at these times. Farmers could pass on that knowledge to the next generation in a way that was easy to remember and unchanging.

Within the constellations, the stars themselves do not deviate from that pattern over our lifetime. However, the overall patchwork of constellations moves as one across the sky over the course of the year as the Earth orbits around the Sun. So the constellations you can see on any given night will depend on the time of year (with the exception of circumpolar constellations like Ursa Major that can sometimes be seen all year round). Often they are combined into seasonal groups – you'll sometimes hear people refer to 'summer' or 'winter' constellations. (In this book I've broken it down even further by organising the night-sky guides into six bi-monthly sections, January–February, March–April, etc.)

Like most of the constellations used by us in the West today, the naming of Ursa Major, which is Latin for the greater or larger she-bear, can be traced to Graeco-Roman mythology. In many ways, I often wonder now why I didn't grasp these stories a little better. They are memorable and can allow us some freedom to give meaning to celestial objects that go well beyond our understanding.

The myth behind Ursa Major originates, like many myths, from the god Zeus's adulterous ways. Zeus had hatched a plan to seduce young Callisto, a beautiful huntress under the guidance of Artemis, goddess of hunting. Disguised as Artemis, Zeus seduced Callisto one morning in the forest whilst she was resting. Callisto fell pregnant, and when this was discovered she was harshly cast out by Artemis for breaking her vow of chastity. When Callisto eventually gave birth to a son, Arcas, her problems multiplied as Zeus's wife Hera learned of her husband's infidelity. Enraged, she turned poor Callisto into a bear who was forced to roam in the wilderness alone.

The story concludes fifteen years later, when Callisto accidently met her long-lost son in the woods. Not recognising her, Arcas was about to shoot his mother with an arrow when Zeus finally intervened. To save them, he transformed Arcas into a small bear and then grabbed them both by their tails and launched them into the sky, where they would live forever in the stars as Ursa Major (Great Bear), and Ursa Minor (Little Bear), another constellation we will explore later in the book.

Some people joke that this story explains why both constellations appear to have long tails when earthly bears do not – they were stretched when they were thrown aloft. The inescapable laws of physics do seem to be used when it fits. This is how I would get off as a teenager. I remember laughing at one of the descriptions of the formation of the Milky Way in particular: that Jupiter put the infant Perseus onto Andromeda's breast, who awoke as poor Perseus was sleeping and shoved him off. The milk spewed out and formed the Milky Way, which for a boy was probably more amusing than it should have been.

As I was growing up, learning about these myths was the easy bit; actually spotting individual stars and finding a constellation was much harder. After some more reading and trial and error as a teenager, I came up with the following approach. Much of the science I've learned in subsequent years, but this routine began my personal journey to the stars, and I hope it may be of use to you too.

This short exercise can be performed wherever you are, in the middle of the city or in the darkest sky environments. No need for a telescope or a pair of binoculars, just your eyes for now, and your curiosity. You can attempt it with your family, your friends or on your own; incidentally the kids will probably be better at it than you. I will explain why later.

Once the Sun has set, head outside. As I grew bigger as a teenager, I became a little bolder and found myself talking to the smaller lads about astronomy and telling them that the clouds weren't really orange; that it was the street lights that were causing the light pollution. The ideal night to see stars is a cloudless one, when there is no Moon in the sky. If the Moon is in its full phase, for example, it reflects sunlight back to Earth and the effect is that the sky gets a lot lighter and washes out fainter stars. It is the largest celestial light polluter, so no or as little Moon as possible produces the best results. The next step is to find north.

If you are stargazing in an inner city, you need to have a reasonable northern horizon, that is you need to be able to see north without your line of sight hitting a building or an office block. To find north use a compass, or if you don't have one try to remember which part of your local sky the Sun set in. Wherever it was, put that part of the sky on your left shoulder, and you are now looking approximately in the right direction. I had lots of problems with this part; I was lost quite often, but what saved me was the sea. The sea? Yes, in Sunderland we are very fortunate to be on the coast. I knew the sea was east and in Grindon I knew which direction the water was. That was my starting point. You may be able to devise reference points of your own.

Along with a clear sky, another key to successful 'naked-eye' astronomy, as it is known, is ensuring the sensitivity of your eyes to dim light is at its maximum. This is known as 'dark adaptation'. The first thing that happens when you move from a bright area to a dark one is your pupils begin to widen to let in as much light as possible. At their widest they are around 7 millimetres across (this will be important when we talk later in the book about binoculars). Next, the cells in your retina called rods and cones slowly become more sensitive to light. This is caused by the action of two chemicals called rhodopsin and iodopsin, which only work in dark conditions. Within six or seven minutes the cones reach their peak sensitivity. However, it takes at least half an hour for the rods to become almost completely dark adapted and they are responsible for most of your night vision. As we get older, the muscles which dilate and contract our pupils get weaker, meaning that for many people their dark-adapted pupil will only widen to 5mm. This means that if you are stargazing with your kids for this exercise, there is a strong chance that they might see and count more stars than you. Their larger pupils will simply let in more light.

Dark adaptation can be ruined in an instant, however. A sudden bright light will quickly reverse the effects of the two chemicals and your superior night vision will be lost. You'll need to start again and let your eyes readjust. You can see why astronomers get angry when someone shines a torch in their eyes or a careless driver beams their headlights across a dark sky. Dark adaptation is important but certainly not essential for this exercise, due to Ursa Major being one of the brightest constellations in the sky. However, I would recommend spending at least fifteen minutes outside to get used to the conditions.

The star chart of Ursa Major (below) is a map of a portion of the sky similar to the one I used as a teenager. It shows the positions of the constellation in the sky as you look north and it also names the stars so that you can memorise and learn them. Star charts are like projections of the sky onto paper and are used by amateurs and professionals to navigate the sky, hopping from one star to another, and after every completed hop learning a little more.

You will need a torch to read this star chart outside, and a red torch will be better as red light has far less impact on our eye adaptation than white. These can be bought inexpensively at most good hardware shops, or you can cover a normal torch with a red plastic film. I wish I'd known this when I was a lad. I used a torch I owned with big 1.5v batteries by Eveready. It threw

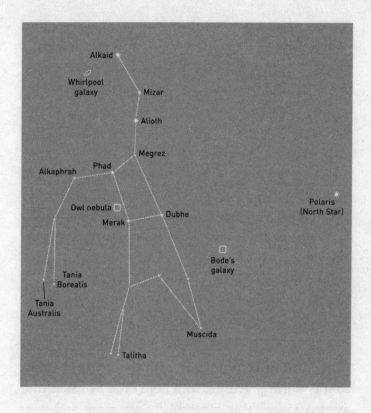

out a lot of white light that didn't help my eyes at all, so I wouldn't recommend that; a red torch will work much better.

Facing north, take a good look up and hopefully with a little practice you will be able to see the constellation. The Plough (or the saucepan) asterism is often the easiest part of the constellation to find first, so we will concentrate on that initially. You're looking for the four stars that make up the 'bucket' of the saucepan – Dubhe, Merak, Phad and Megrez. You can then skip down the three stars which form the 'handle' of the pan – Alioth, Mizar and Alkaid. Once you have found this, try looking directly at Mizar. Can you see a faint companion star next to it? This star is called Alcor. If you are struggling to make it out head-on, try looking at it out of the corner of your eye. It should appear brighter. This is known as averted vision, and is a great trick that works on all celestial objects. It happens because of the differences between the rods and cones in your eye. Your cones are mostly clustered on the centre of your retina, whereas the rods are denser on the edges. So if you are ever struggling to see a faint light, looking out of the corner of your eyes allows the rods a greater chance of picking up the dim light. Depending on the strength of each eye, it makes a difference which side you avert to, so it is worth practising with both eyes. By using your averted vision, you can actually see more and your ability to resolve any star increases.

My next tip is to familiarise yourself with the constellation. Try to recite the names of the stars, or to remember their shapes. Are there any common features between the stars; maybe some are brighter than others? Do others appear smaller or bluer, or maybe some appear to have fainter companions very close by. Sketching what you see can be a good way of trying to remember how to spot them, or their differences. During my teens I started keeping an observation log in a small notebook, which I keep to this day as a simple diary of what I see when I go stargazing. I record the date, the weather conditions and the temperature, which can help improve future sessions, and I write down memorable features that I see, from shooting stars to the hue of stars, from a vibrant blue to a punchy yellow. Later, if you take up astrophotography, you can take photos to accompany your notes. As a boy, these tips systematically developed my understanding of how to observe.

The only nuisance I experienced growing up was that Ursa Major, like all stars and constellations, appeared to move across the sky each night, drifting from east to west over the course of the evening. This made rediscovering it tricky as the night wore on. I had read that the stars themselves are so mind-bogglingly far away that their actual motions are hardly discernible. I resolved this as a boy by thinking of them as if I were watching a 747 aircraft take off. If you are close to the airport runway, then boy does it move, but if you see that same aircraft from miles away cruising at altitude, its motion seems much slower. Nevertheless, the stars and constellations do seem to physically move across the sky together each night from our viewpoint. So what is happening?

This apparent motion is caused by the Earth's rotation and its orbit around the Sun. Because our home planet spins happily on its axis, once a day, every day, and because we are perched on this spinning ball, our view of the sky above is also one that spins, constantly. This creates the apparent movement of the constellations – but it is we that are actually moving, not the stars.

Having understood this, I then realised that I needed another aide-memoire that would be more reliable in order to find particular constellations whenever I wanted to. Thankfully I was in luck. The Moon may have been my first light, but Polaris, or the North Star, was my most important discovery. A light that I, and countless astronomers and others before me, have used as a jumping-off point to navigate the night sky.

Polaris is significant because its position never appears to move for us in the sky. Why? Well, imagine if we had a huge pole that we passed through the Earth and out of the other side, running from the South Pole to the North Pole. The Earth would rotate on that pole, representing the Earth's rotational axis. If you extended that imaginary pole beyond the North Pole into space, you would eventually reach the star Polaris, which happens to sit almost directly above our axis of rotation. Because of Polaris's location, it barely seems to us to move at all. Another way to think of it is to imagine standing in a room with a cross marked on the ceiling above your head. If you spin on the spot, the room will seem to you to move, but the position of the cross will never deviate. Polaris is the equivalent of that cross – it is always within a degree of due north, a fact that sailors and other navigators have exploited for centuries to find their way. Walk towards Polaris and you are walking north; get Polaris on your left shoulder and you are walking east and so on.

For us stargazers in the northern hemisphere, the effect of Polaris's position north of the Earth's rotational axis means that all other stars and constellations appear to spin around Polaris, so that if you find Polaris, you can reliably jump to any other constellation.

Locating Polaris for yourself is straightforward with the use of a star chart (you can use the ones in this book). Start with the seven stars of the Plough in Ursa Major. Then find Dubhe and Merak – the two stars furthest from the handle in the bowl of the saucepan. If you follow the imaginary line between them and extend it out from Dubhe in the same direction, you will be carried straight to Polaris. For this reason Dubhe and Merak are often known as 'The Pointers'.

Most people have a common misconception about Polaris, that it is the brightest star in the sky. That accolade actually goes to Sirius, Greek for 'glowing' or 'scorcher', which can be found in the Canis Major constellation. Polaris sits in the constellation Ursa Minor, where the mother–son bond between Callisto and Arcas is alive and well. However, Polaris is undeniably the most important to amateur astronomers. And to me.

*

Around the time I left school, Polaris was one of the few constants in my life. In 1982 the dole queues were long. My first stint in line lasted for six months until I got a chance with the YOP, a Youth Opportunities Programme, working for British Telecom. I was employed to observe and learn, but I was hoping I would get a job there on a permanent basis. It wasn't to be. I was soon back in the dole queue again. Like some sort of prisoner I remember being summoned to the counter and interrogated at my allotted time. 'Have you looked for a job? Where? When?' Give me a job and I will do it, I wanted to say. I felt hopeless.

The climate in those years can only be described as grim. Maggie Thatcher was dismantling our region, the Falklands War was raging and opportunities were generally sparse. I felt like we were being stripped of everything, not least the Port of Sunderland on the river Wear, which runs through the heart of the city. Once famed for churning out great ships of every kind, our yards could not be matched by any other part in the world for skill and dedication. Doxford's, Austin and Pickersgill – all were household names to us living on Wearside. When I was very young I used to ride my bike down to the docks at weekends and watch the fishermen coming in on shining boats, some with sky-high yellow cranes. They carried plentiful crates of rainbow-coloured fish and crabs. But by the time I left school it was a way of life that was dying. Soon, smaller fishing boats were moored at the dock and were the only ones that would chug out on a daily basis. I guess it was still a working life for many then, but silently in the background the vast empty yards that had once made this city famous looked on. They were becoming huge empty shells.

It was the fixed point in the sky that rescued me during that time. As my work life narrowed, my passion for astronomy blossomed. With Polaris as my guide I began to explore the rest of the universe like a celestial dot-to-dot, navigating from constellation to constellation, wherever I was and whatever the season. My skills improved and I now used my telescope, the same one that I had borrowed at Christmas from my brother eight years before, to find deeper-sky objects like gas clouds and even faraway galaxies. My first light of the Moon was a memory; my future lay in the light of distant suns. But a period of darkness still lay ahead.

January–February Sky Guide

Orion–Taurus–Canis Major–Auriga–The Moon

ORION

In winter (January) looking south from Kielder Observatory

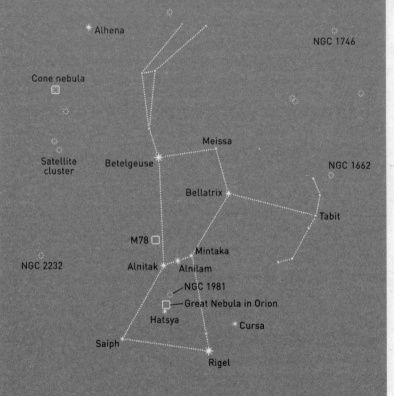

Alhena

NGC 1746

Cone nebula

Satellite cluster

Meissa

Betelgeuse

NGC 1662

Bellatrix

Tabit

M78

Mintaka

NGC 2232

Alnitak

Alnilam

NGC 1981

Great Nebula in Orion

Hatsya

Cursa

Saiph

Rigel

Mirzam

Stars   Mag 0 ✷   Mag 1 ✷   Mag 2 ✶   Mag 3 ✦   Mag 4 ·   Mag 5 ·   Clusters ✧   Nebulas ▢

Orion is one of the most well known and splendid of the constellations. From the northern hemisphere it is easy to spot high in the southern sky on clear winter nights, and its bright 'Belt' stars, which twinkle and flicker a multitude of blues, yellows and oranges, are where you should start as your points of reference.

There are many legends in Greek and Roman mythology about Orion. The son of the sea-god Neptune and the Amazonian huntress Queen Euryale, Orion was an exceptional hunter and bragged he could kill any animal on Earth, until he was promptly killed by a scorpion. Zeus intervened and placed Orion and Scorpius as far apart in the sky as possible to ensure they were never visible at the same time again.

During the autumn months in the northern hemisphere, you will have to wait until the early hours of the morning to catch a glimpse of him. As the year progresses, however, he rises earlier and earlier, until by the time the new year rolls around, he rises just as the Sun sets. By 10.30 p.m. on 1 January he will appear due south – the area astronomers refer to as the 'Theatre of the Sky' – and is primed for viewing. By the end of February he's there as early as 7 p.m.

Orion is easily recognised by the three stars – Alnitak, Alnilam and Mintaka – that make up the famous Orion's

Belt around his midriff. Equally easy to spot are the four stars which frame them – his shoulders and feet. Two of these stars – the red supergiant Betelgeuse (left shoulder) and the blue supergiant Rigel (right foot) – are particularly gleaming, both ranking in the top ten brightest stars in the sky. Betelgeuse, which is said to mean 'armpit of the Central One' in Arabic, is a star on the brink of demise. Some time in the next few million years, it will explode as a spectacular supernova and be so bright that it will, for a time, be visible during the day.

Across his body, Orion's right shoulder, Bellatrix, is brighter than any of the Belt stars, whereas Saiph (left foot) is about on a par with them. Many illustrations of Orion show him with his club and shield. The club extends north from Betelgeuse, the shield west from Bellatrix.

Under dark skies, you will easily spot the Sword of Orion hanging down from his belt. At first it might seem like three stars in a line, a vertical version of Orion's Belt. However, look more closely and you'll notice that the central one of the three 'stars' has a natural fuzziness about it, and binoculars and telescopes are certain to reveal its true identity.

The Sword of Orion is the Great Nebula in Orion (M42), a giant cloud of gas and dust. If Orion is the most famous constellation, then M42 is probably the most famous deep-sky object in astronomy. The 'M' stands for the surname of an eighteenth-century French astronomer called Charles Messier who, from his lofty apartment in Paris, would sweep the sky looking for comets. As he did so, he stumbled across some other fuzzy objects that he couldn't explain, but which clearly weren't comets at all. He listed them and formulated a table of these 'M' objects, all 110 of them, to be avoided by future comet-hunters. However, once technology got better and these 'M' objects were revisited, they turned out to be some of the most beautiful of celestial objects, from gas clouds to distant galaxies. They are now among the most viewed and popular objects in amateur astronomy.

Sitting around 1,500 light years away from us, M42 is a star factory, a stellar nursery where infant stars are born. Gravity is slowly pulling this material together to form pockets of gas in which the temperature and pressure increase enough for brand new stars to ignite out of the darkness. At its heart lie four stars known as the Trapezium cluster. Can you spot them? They will appear as four tiny points of light right at the centre of the brightest region and they can be glimpsed with a big pair of binoculars or a small telescope. Radiation produced by these stars causes the surrounding gas to glow and this is how we see the gas cloud (nebula).

Peering into the murk with the Hubble Space Telescope

has revealed dark, flattened discs surrounding some of the newborn stars. Known as proto-planetary discs, or proplyds, these stars represent the first stages of planetary formation. Gravity will eventually sculpt the material in these bands into brand new worlds. Astronomers believe our solar system was formed in a very similar fashion some 5,000 million years ago. So in many ways this region that you will be able to see, or even be looking at now, is a star-forming region where new solar systems are being formed, right before your eyes. It is a seductive thought that we can observe such vast acts of nature from our own backyards.

Another very popular target in Orion is the Horse-head Nebula, although it's only visible from the darkest skies and you may need a filter for your telescope (filters can improve detail by blocking a part of the colour spectrum). Resembling a knight piece from the chess-board, it is a lot smaller and dimmer than its neighbour and so represents a challenge for the beginner. It has been made famous by spectacular long-exposure images from professional telescopes, and a sizeable aperture is certainly required if you're going to see it for yourself. To find it, start with the first star in the Belt, Alnitak, and head off towards Saiph (Orion's right foot, opposite Rigel). You aren't travelling far, just around half a degree (about the width of the Full Moon). If you reach the faint star HIP 26820, then you have gone too far. Its distinctive equine silhouette is the result of its dust and gas blocking the bright, pink light from an intensely glowing cloud of hydrogen behind it.

You could also turn your attention to the Flame Nebula just to the left of Alnitak. Much of the glow of this gas cloud is caused by Alnitak casting intense radiation into the region. Alternatively, just to the left of the line joining Alnitak and Betelgeuse is the nebula M78. Discovered in 1780, it should be easily visible through small telescopes. In particular, you'll see two tenth magnitude stars whose light we can thank for allowing us to see the surrounding gas.

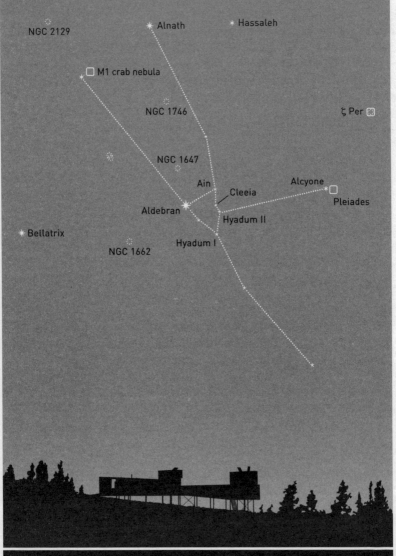

TAURUS

In winter (January) looking south from Kielder Observatory

NGC 2129

Alnath

Hassaleh

M1 crab nebula

NGC 1746

ζ Per

NGC 1647

Ain

Alcyone

Cleeia

Aldebran

Hyadum II

Pleiades

Bellatrix

Hyadum I

NGC 1662

Stars   Mag 0 ✸   Mag 1 ✴   Mag 2 ✶   Mag 3 ⋆   Mag 4 •   Mag 5 ·   Clusters ❋   Nebulas ▢

Once you have located Orion, finding Taurus – one of the zodiacal constellations – is relatively straightforward. Start with the three stars in Orion's Belt and draw an imaginary line linking them together. Now extend this line upwards and towards the right, skirting underneath Bellatrix and through the shield of Orion. Very quickly you will arrive at Aldebaran, a yellow-orange star that is the brightest in Taurus and the fourteenth brightest star in the heavens.

Informally referred to as the 'Bull's Eye', it sits at the bottom left branch of a V-shaped group of stars known as the Hyades cluster which make up the head of the bull. It may look as if Aldebaran is part of this group, but it isn't. Instead, it is much closer to us than the other stars, meaning its apparent membership of the cluster is simply an optical illusion. The Hyades itself is the nearest open star cluster to our own solar system and therefore has been studied at great length by professional astronomers.

If you extend both sides of the V-shape upwards, you will encounter the stars that make up Taurus's horns. Beginning at the base of the V and moving down will lead you instead to the bull's feet. The main chunk of his body lies to the right of both of these areas, and it is here, in the shoulder of the bull, that you will find one of the most famous and popular deep-sky objects in astronomy.

As with a number of the constellations, Taurus is just Zeus in disguise. Famed for his wandering eye, Zeus would often covet mortals despite the fact he was married. On one occasion, the subject of his affections was Europa, a princess of Phoenicia. In order to woo her, he took the form of a magnificent white bull and approached her whilst she was strolling along a beach. So enamoured was she by the creature's beauty and playful nature that she climbed on its back. Zeus responded by suddenly whisking her off into the sea, later to surface on the island of Crete, where he revealed his true identity and seduced her. In the sky, only his head, shoulders and legs are visible – the explanation often given is that his hindquarters are submerged in the water.

Just like people, the stars we see in the sky cover a whole range of ages. The stars that make up the Pleiades cluster (M45) are some of the youngest at just 100 million years old. Find them by following Orion's Belt straight through the V-shaped Hyades cluster.

Nicknamed the Seven Sisters, after the mythical siblings that Orion tried unsuccessfully to marry, the cluster actually contains over a thousand stars. They were all born together out of the same cloud of gas and dust, in very much the same way as the stars currently being manufactured in the nearby Orion Nebula. An age of 100 million years may sound extremely ancient when considered in human terms, but for most stars that is very young indeed. Take our Sun for instance – it is 5,000 million years old and is almost exactly middle-aged. So if our Sun were human, it would be around forty years old. The Pleiades stars would be just nine months old.

In terms of observing it for yourself, perhaps binoculars are best. However, if you want to use a telescope, then you will need a wide field eyepiece and a short focal ratio instrument, meaning you'll only ever get to see part of the cluster at a time. Whereas, with binoculars, several dozen of the brightest stars are nicely framed in a single field of view and you can happily drink in the entire spectacle. Look for the faint blue glow that envelops this area; this is a cloud of gas and dust that the cluster is passing through. Because of the UV radiation flowing out of these young stars, the gas and dust scatters and reflects the starlight, which is known as a reflection nebula. This blue example has to be one of my favourite objects as it is easily recognisable in the winter skies.

The other grand deep-sky object in Taurus is the Crab Nebula (M1), located very close to the star marking the tip of the Bull's southern horn. Unlike the Orion Nebula, which represents an area of star birth, M1 is the remnant of a star's catastrophic demise. In the year 1054, Chinese astronomers recorded the sudden appearance of a 'guest star', one that shone so bright it could even be seen during the day. We now know that they saw a supernova and that the Crab Nebula we see today is the shrapnel left over from the blast.

It is only big stars – those with a mass more than eight times that of our own Sun – which explode as supernovae at the end of their days. So, as we'll see in the pages to come, another fate awaits our own star. The cores of those large stars eventually collapse and the rest of the material in the star begins to collapse inwards too. The collapsing core creates a giant shockwave which speeds outwards, tearing through the in-falling material and causing an almighty explosion – so mighty that the resulting light can outshine all the stars in a galaxy and be visible in daylight.

If you're keen to take a look at the nebula, then you're likely to need a telescope. Even a 4-inch telescope should begin to reveal that some parts of the cloud are brighter than others. It will appear as a faint glow of fuzz with a slightly elliptical appearance. These objects are the remnants of one of the most energetic events in the universe, without which we couldn't exist, as these events are responsible for the production of heavy elements.

CANIS MAJOR

In winter (January) looking south from Kielder Observatory

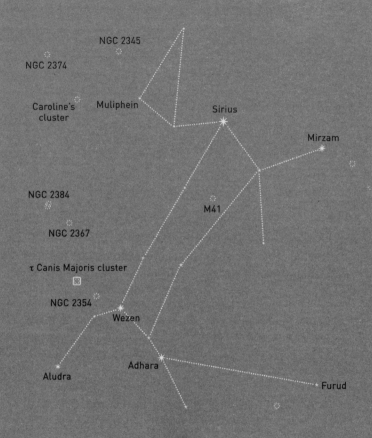

NGC 2345

NGC 2374

Caroline's cluster

Muliphein

Sirius

Mirzam

NGC 2384

M41

NGC 2367

τ Canis Majoris cluster

NGC 2354

Wezen

Aludra

Adhara

Furud

Stars   Mag 0 ✸   Mag 1 ✴   Mag 2 ✦   Mag 3 ★   Mag 4 ·   Mag 5 ·   Clusters ⬡   Nebulas ☐

In most star stories, Canis Major is Orion's chief hunting dog and faithfully follows him across the sky year after year. Many suggest he is perpetually hunting Lepus, the Hare, a small constellation found beneath Orion's feet.

Canis Major is absolutely dominated by one star in particular – Sirius. Known colloquially as the Dog Star, it is the brightest star in the night sky in the northern hemisphere. In fact, it derives its name from this appearance as it comes from the Greek word for 'searing' or 'scorcher'. It owes its dazzling luminosity to the fact that it is one of the closest stars to the Earth. Locating it is easy – just extend the line between the three Belt stars of Orion down and to the left as you look at it.

One of the first things you'll notice is that it twinkles very violently, often flitting between one colour and another. It will mostly alternate between red and blue, but other colours, including green, can sometimes be seen. Whilst it may appear to be a single star, it is actually part of a binary pair with a star called Sirius B (the main star is regarded as Sirius A). However, this companion appears 10,000 times fainter to us.

None of the other stars in the constellation comes close to Sirius's brightness; however, three are above magnitude 2.0 and include Wezen (at the base of its tail), Adhara (top of the hind leg) and Mirzam (front paw). Yet arguably the most famous star in Canis Major after Sirius is VY Canis Majóris. Too faint to see with the unaided eye, it was once the biggest star known to astronomers. Were it to replace the Sun in our solar system, its outer edge would surge all the way to the orbit of Jupiter.

As the band of the Milky Way passes directly through this constellation, there are a few open clusters available to see here. The most notable is M41, which you can find with binoculars by starting with Sirius and tracking around 4 degrees (eight Full Moon widths) due south into the body of the Dog. There are around 100 stars in the cluster, all within an area of sky equivalent in size to the Full Moon. Like the Pleiades in Taurus, they are very young stars with an estimated age of 190 million years. The cluster is particularly attractive because the stars are markedly different colours, with blue, yellow and orange members.

Another open cluster, NGC 2360, can be found by starting with Sirius, heading east and dipping just below Muliphein (the star at the top of the Dog's head). You'll also hear it called Caroline's cluster after its discoverer, Caroline Herschel (sister of the discoverer of Uranus, William Herschel). She was paid to observe the stars by King George III, making her the world's first paid female scientist.

Observers with sizeable telescopes may also be able to make out the merging spiral galaxies of NGC 2207 and IC 2163. Begin with Mirzam and head just to the left of the fifth magnitude star HIP 29941. On an unswerving collision course, it is thought they will eventually combine with one another in around 1 billion years. They were discovered in 1835 by John Herschel, William's son and Caroline's nephew.

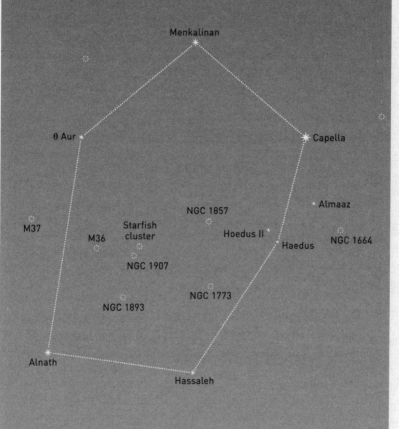

AURIGA

In winter (January) looking east from Kielder Observatory

Menkalinan

θ Aur

Capella

Almaaz

NGC 1857

M37

Starfish cluster

Hoedus II

M36

Haedus

NGC 1664

NGC 1907

NGC 1773

NGC 1893

Alnath

Hassaleh

Stars   Mag 0 ✸   Mag 1 ✦   Mag 2 ★   Mag 3 ⋆   Mag 4 ·   Mag 5 ·   Clusters ⋰   Nebulas ☐

Auriga represents a legendary charioteer. In Greek myths he is often thought to have been Erichthonius of Athens, a hero born of the soil and raised by the goddess Athena. It is said he invented a chariot drawn by four horses, which he used to great effect in battle to overthrow Amphictyon and become the king of Athens.

To track down Auriga, it pays to start, once more, with Orion. Locate Alnilam, the middle star in Orion's Belt, and move upwards, passing halfway between Betelgeuse and Bellatrix and skipping by the star Meissa in the hunter's head. Continuing upwards through the very tips of Taurus's horns will take you into Auriga, which looks like a squashed hexagon. Due to its slightly higher location in the sky, the charioteer is circumpolar – it never sets for observers above a latitude of 44°N. However, it is best seen in the first few months of the year.

Upon reaching Auriga you'll quickly notice the yellow-hued Capella, the third brightest star in the northern hemisphere. Like Sirius, Capella is actually a multiple star system. Two bright giant stars – Capella Aa and Capella Ab – not only orbit each other every 109 days, but in turn orbit a separate pair of red dwarf stars.

Capella gets its name from the Greek for 'small female

goat' as the charioteer is often depicted holding the animal. Hence it is sometimes called the 'Goat Star'. The farmyard theme continues if you look to the lower right-hand side of Capella and find the three fainter stars – Epsilon, Zeta and Eta Aurigae. Arranged in a small triangle, they are affectionately known as 'The Kids'.

To the lower left of Capella you'll spot the star Menkalinan and at the base of the squashed hexagon sits Alnath, a star that was once regarded as part of Auriga but is now officially considered to be in neighbouring Taurus.

Just like Orion, Canis Major and Taurus, the Milky Way also passes through Auriga. However, this constellation sits 180 degrees away from the galactic centre. So looking this way means we have our back to that crowded, dusty region. For this reason the density of background objects is lower here.

That said, there are three Messier-listed open clusters for you to hunt down: M36, M37 and M38. They were discovered by Italian astronomer Giovanni Battista Hodierna before 1654 and can all be seen with a pair of binoculars. Start with the star Alnath (the one that's technically in Taurus) and draw a line up and slightly left towards the star Theta Aurigae. The three clusters sit on either side of the rough midpoint of this line – M37 to the left of it and the other two to the right.

M37 is the richest of the trio, boasting over 500 stars. It was once described as 'a virtual cloud of glittering stars' and is thought to be around 500 million years old. Skip across that imaginary line to M36 and you'll see a cluster thought to be very similar to the Pleiades in Taurus but less visually impressive because it is almost ten times further away from us. About forty of its stars will be visible to you and it's the smallest of the three. If you edge across to M38, the most diffuse of the three clusters, you'll see its brightest stars resembling the symbol for the Greek letter Pi (π). These three clusters are a treat on clear moonless nights; they really come alive. I always visit them when explaining star clusters, as they are so easy to see.

The Moon

The Moon can be viewed any time of year and is one of the best celestial objects to observe in close detail, particularly if you are new to astronomy. This is because it is by far the nearest natural object to us in space at an average distance of only 239,000 miles (384,000 kilometres). Due to its proximity and brightness, you need very little magnifying power to glimpse the vast array of treasures the Moon has to offer.

First you'll need to get to grips with its phases. The amount of sunlight the Moon reflects towards us changes as it orbits around the Earth. It takes just shy of a month for these phases to be completed. In fact, that's where the word 'month' comes from – it used to be spelled 'moonth' and represents the approximate time it takes for the phases of the Moon to repeat.

Newcomers to astronomy are likely to assume, quite naturally, that the best time to observe the Moon is when it's full. However, a Full Moon is a gigantic pest for astronomers. Its harsh light washes out faint nebulae and galaxies, and looking at the Moon directly at this time is little consolation – there are better times to view it. The very reason the Moon shines is because it is continually reflecting light from our Sun.

A particularly good time for observing the Moon with binoculars or a telescope is when there is a distinct dividing line between the light region and dark region (such as at First Quarter when it is half illuminated). Known as 'The Terminator', this line is where lunar day meets lunar night. And if you think about dawn and dusk on Earth, shadows are much longer at these times. These shadows lend a wonderful perspective to the Moon's litany of craters, mountains and volcanic ridges that you just don't get at Full Moon.

Due to its lack of atmosphere, active volcanoes or flowing water, there is no mechanism for wearing away the craters – scars from the myriad impacts the Moon has sustained over billions of years. One of the youngest, and most notable, is Tycho. Seen from Earth's northern hemisphere, it is located towards the bottom of the Moon and is easily recognised by a series of bright lines or 'rays' spreading outwards from it like the spokes of a bicycle wheel.

As the impactor that created Tycho slammed into the Moon, it melted some of the rock and threw it into the air. Hitting the icy coldness of space, these lava droplets soon solidified and fell back to the lunar surface as glass beads. Now they glint in the sunlight like cat's eyes. Not far from Tycho is the unmistakable Clavius, the third largest crater visible. Here four smaller craters curve in ascending size order like a smile inside a larger, older crater.

The Moon's craters are superimposed onto a system of very large, dark patches known as 'maria' or 'seas' which make up the infamous 'Man in the Moon' face. Astronomers of yesteryear believed they contained water, but today we know they are ancient dried-up lava plains. The most famous is the Sea of Tranquillity where Neil Armstrong first set foot on the Moon. However, there are many others, including the Sea of Nectar, the Sea of Cleverness and the Sea of Crises.

Despite Pink Floyd claiming otherwise, there is no

permanently dark side of the Moon. The same side of the Moon always faces the Earth and we see the phases of the Moon because the illumination of that face changes as the Moon orbits us. At Full Moon it is completely lit up. Conversely, at New Moon, we don't see any light as it is the far side that is illuminated.

It is this opposing face that is often labelled 'the dark side'. However, the only time that it is completely dark is during our Full Moon.

2

Dark Skies

13 February 1985.

Sunderland is playing Chelsea in the semi-final of
the Milk Cup. Win this and we are on our way to
Wembley. Memories of the 1973 FA Cup Final, when
Jimmy Montgomery's miraculous save helped us beat
Leeds United 1–0, bond the city together once more.
We dream. Only the Cockneys stand in our way now.

It's around lunchtime when we make our way into
town. We meet in a corner pub called the Londonderry,
a faded majestic Edwardian building topped with copper
domes, situated just twenty minutes away from the
ground. It's packed inside as usual when we arrive: a
throng of football lads conspicuous against the polished
timber and deep red carpets. Not in football shirts, mind.
The dress code for our group is always 'no colours'.

I spot five of the lads at the end of the long wooden
bar and wander across to meet them. I'm with my

younger brother Marty, who breaks off to join a poker game in the corner by the piano. I order a lager.

'Ow, Gaz,' comes a deep voice from one of the tables.

'Areet,' I reply. I don't know him but I recognise his pock-cheeked face.

'You lads all set?'

I look at him and nod.

'*Always*, mate.'

This is how it goes. A little probe from someone in the know.

Conversation turns to the Cockneys. We're used to the London teams coming up north in numbers on their plush coaches, waving fivers around, all too aware of the plight of our closed pits and rusted shipyards. They wear the latest designer clothes: Sergio Tacchini, Lacoste and Pringle. A stark contrast to our donkey jackets, hand-me-downs or, worse, my dreaded Geordie jeans – faded and tight in all the wrong places. But Chelsea are the worst. They rub everything in our faces and today they're travelling in force. There might be 15,000 of them.

I'm interrupted by the noise of glasses clattering on the table behind us. It's Marty. I turn around and he's squaring up over the cards game, an ashtray crashing to the ground. Through the smoke he looks too young to be here – his sandy blond hair and fresh face betraying his age. He's just eighteen. But he's giving as good as he gets.

'Ya fucking did, man!' yells his accuser, a wiry older lad, veins pulsing at his temple.

'Nor I never, man,' Marty shouts back, balling his fists and leaning in to the man's face.

I think to myself that if I know Marty, whatever it is, I'm sure he's done it. But the other lad is older. It's my duty. I walk over slowly.

'It's Fildesies' brother, man,' I hear someone say in a quieter voice.

'I dinnit care who he is,' snarls back my brother's enemy. My enemy now.

'What ya say, mate?'

I'm interested. I stand up behind my brother, eking every inch out of my 6 foot 2-inch frame. A full head taller than him. I square out my shoulders as well, trying to show that this lad doesn't scare me one bit. I feel terrified inside.

'He's cheating, man.'

'So what!' I fire back.

We glare at each other and the group goes silent. They know as well as I do what is at stake here. I walk closer to the lad. I put my beer down. He's more sheepish now.

'Nowt, I didn't say nowt.'

'Well, keep ya mouth shut then, eh?'

Marty is buzzing as we leave the Londonderry. 'Giddin, Gaz lad!' The fact that he was certainly cheating is irrelevant. We walk down to the Central pub to meet the rest of our mates. A poky place with closed-in walls, it's a gay bar during the week, but on match days it's our base. I spot one of my best friends, Mick Gibbon. He may have a small frame but he's the gamest lad I know. Mick is a welder and pipe-fitter by trade, and he doesn't stand down to anyone. Today he is decked out in smart jeans and a polo shirt, the latest fashion fresh out of Manchester – he's on good money.

'Ow, Mick lad,' I say as we hug.

'Ow, Gaz,' all the lads sing back in unison. I feel elated that my mates are together again.

'This is it,' I say to Mick, starting to feel giddy with excitement.

'You better believe it, Gaz.'

'Any sign of that lot?' I ask. As usual, we're relying on the grapevine for news of 'the meet'. Plans have been known to get a bit lost in translation. If someone says there are fifty lads, there can often be only ten. But it also works the other way.

'I've heard they're through Washington and down Seaburn,' Mick speculates. It seems the Chelsea fans are staying out of the city centre for now. 'Been a few scraps near the train station though. Shall we have a wander?'

'Nah,' I say. 'Let's just stay here for now. See what happens.'

Over the next few hours, famous songs like 'Sunderland till I die' or 'We hate Cockneys' ring out in the pub like battle cries, made louder by the echoes off the tight walls. It feels hot and humid. Sweating bodies sway and half-filled glasses of lager are raised aloft. Some of the tables have been pushed together and a few of the lads are standing up on them, now leading us all in song like a choir. The air stinks of cigarette smoke, bad breath and stale lager, but morale is building and squeezed in among the thin walls the pressure of expectation is rising too. We are waiting for the trigger, that single thing to ignite the touchpaper and set it off. We don't have to wait long.

Shouting rattles the windows of the pub. Chelsea supporters are outside. Around me men dart for the single door to get out and fight, desperately clawing to squeeze through the narrow door frame. We stumble into the light of Fawcett Street, greeted by the sight of a battlefield. Hundreds of fans from both sides swarm the road leading to the bridge over the river Wear. A melee of arms and legs on the ground: punches and head-butts, bodies writhing, southern and northern accents swearing and screaming, pain and exhilaration. I have to get some, so in I go. I throw a flurry of punches at the nearest man I see, a skinhead in a yellow jacket dressed far too smartly for our group. He falls to the ground and shrinks away, but then I feel a kick at my leg. I turn around to be tackled to the ground by a policeman. Desperately trying to wriggle free, I scan the throng for an ally. I see Mick close by. He pulls my shoulders and helps me escape from the copper before we sprint away from the ruckus to regroup with Marty and our mates. We need to conserve our energy – this is just the warm-up before the main event. Shirts torn and skin bruised, the adrenaline is flowing and we sing our heads off as we move towards the bridge. The twenty of us are laughing and breathing heavily. We join a larger group led by a police escort in riot gear and follow 200 or 300 Chelsea fans in similar numbers to our own. As we approach the bridge I can see other fights breaking out in front: a sea of heads and men running around in all directions over the bridge. I look at my brother. My brothers. I am twenty years old. I feel alive.

*

The fighting began around the time I left school. It was the early 1980s and the government's plan to rid the area of heavy industry was in full swing. The order books dried up and the shipyards were closing down. Slowly and methodically thousands of jobs were being lost, and the impact was felt across the North-East. I was unemployed and becoming depressed and frustrated. I wasn't the only one – anger hung over the town like a cloud. My astronomy didn't matter for the time being; the stakes were higher. We needed to earn a living. We wanted to vent. Someone, somehow, had to pay.

Like most men of my age, I channelled my frustrations into one of Sunderland's few constants, SAFC.

During my early teens my dad had briefly worked at Roker Park, the home of Sunderland Football Club, which made the obsession take stronger root in me. He was the manager of the merchandise store and I was fortunate to get the odd free match-day ticket. Mostly, however, I would sneak into the game on a Saturday with my Grindon mates or go for the last ten minutes when the gates were open for some of the crowd to leave early. Watching the team and being amidst the fans was the highlight of my week; it was so much better being part of an even bigger gang now, all in red and white stripes. Even the Thornies were on our side as we chanted 'Haway the lads' and 'Sunderland, Sunderland, Sunderlaaand', the songs belted out around Roker Park by the 30,000-strong crowd. I met new mates there who reinforced my conditioning – school and education had been a distraction; we were blue-collar and proud. We lived for football on Saturdays and would put up with a trade or the military during the week. But soon the games themselves weren't enough.

My first taste had been a home game against our arch-rivals Newcastle United. It was routine for a small gang of us, comprising my brother Marty and a few other close friends, to meet for a drink before the match. This time, our group met in the centre of town, along with about a hundred others. It was a sunny day and I remember boys everywhere were dressed in shirts, jeans and Dr Martens boots; none of us were in our club shirts. Shouts rang out and everyone started to run to the east end of town, towards the docks. We were looking for black and white shirts but they were dressed just like us – in their casuals, which separated them from the common Magpie fans. I was amazed at how easy it was to get into a fight, so I dived in, but these were men and I was a boy. I got separated from the lads and ended up in the middle of the Magpies. Two men in their forties held me against a wall by the throat and ripped my red-and-white scarf off me. I ran all the way home without looking back. But I would return soon enough.

*

Sunderland 2–0 Chelsea. We spill out of the stadium in euphoric mood. Thanks to two Colin West spot-kicks we've won the game on the pitch; now we have to win on the streets. Police are patrolling the alleyways and back-roads: on horseback; on foot; some with helmets, wildly lashing out at people with truncheons; others with dogs snarling and snapping, baring their teeth. I see one Chelsea lad pinned down on the ground as a German shepherd chews on his leg; he is screaming but we have to move on. By around 10 p.m. I am still in a sea of Sunderland fans washing slowly through the streets away from the ground. A few of us suddenly notice a smaller group up ahead next to the programme shop, where you could buy small books for a particular match. It is difficult to make out which side is which in the darkness. I am still with my gang and there must be about a hundred of us lads when the cockney voices start to ring out. Bang. They are upon us. These boys are brave and want to fight; we duly oblige. I have to defend myself. It seems like we go toe to toe for hours, lurching and pushing and kicking and scrapping. On the ground and running about. There is just us, no innocent bystanders; just my gang, their gang and the police. It is a free-for-all scrap and again I feel alive. My dad's words, 'Rejoice, oh young man, in thy youth' ring in my ears, the sheer adrenaline racing through my veins. I feel like I am fighting for something, anything, I don't care. I just want to feel that I'm not useless.

I stand victorious under a lamp post after roughing up my latest opponent. I'm not aware of the mounted police charging towards us. I see it momentarily before it hits me. It looked like a policeman throwing his helmet – but that wasn't it. The object hits me straight in the forehead and I fall dazed to the ground. On all fours, I feel blood running warmly down my face. I can see it pooling in the cracks of the tarmac. Voices and footsteps are everywhere, so I try to get up, but I can't. A big hand is on my neck and I hear a voice say, 'Stay down, stay fucking down.' The hand holds me there for what seems like an age but must have been no more than ten minutes. It wasn't a local accent.

Eventually I feel the hand release me and the figure runs away. I stand up gingerly and wait until my mates and Mick Gibbon come back. They help me limp away and we end up in a back lane, with police sirens still blaring nearby. The lads dress my wound with an old woolly hat and finally we set off home. Fights continue around the city in other pockets in what will be long remembered as a dark night, not only for football but for Sunderland itself. Numerous fans are injured and arrested. Later some police will lose their jobs for malpractice.

I am in hospital late into the night getting six stitches to my head wound. I learn that it was a Chelsea fan who had seen the brick hit me in the forehead. He made the decision to protect me until most of the fighting had passed. Towards midnight I am discharged from the ward and set off home to my young wife and babies. I don't feel proud, like I sometimes do after these fights. I feel confused. I am addicted to the thrill and my feelings of comradeship and belonging with the gang are real. But I know I am being selfish and destructive. I am sabotaging my family and I have to stop. My wife Maureen is crying and horrified when I get home. She rings out some home truths about responsibility and what it means to be a father and a husband. She is dead right and I feel shame. I have to change.

*

I met Maureen when I was sixteen years old, playing in town at the local amusements arcade. By the time we were eighteen we were married and had our first son together, Gary Daniel. I moved out of my parents' place and we rented a small house in Pennywell, a run-down part of town in Sunderland. Not long after we had saved for our own place in a better area, an estate called Sunnybrow. In our new home I felt fulfilled and safe; I had re-created the family unit I had drawn so much strength from as I grew up. Only now I was a dad, and I was behaving selfishly. The night I was discharged from hospital, Maureen was furious and disappointed. 'You could at least do what your Neill's doing.'

A friend of mine, who I had gone to school with, had been working as an apprentice bricklayer. He said the money was good and told me they were looking to hire. In my head, the job sounded like a decent fit; it chimed well with my social ideal of being a manual labourer, of wearing a blue collar and just getting on with it. Being a brickie would keep my football mates off my back too. Gazing at stars would not. If anything, I thought it would be a stopgap measure until something more permanent turned up. I enrolled at Wearside Technical College for a six-month course to begin my apprenticeship as a bricklayer.

On my first day, I arrived at a large workshop, completely empty aside from a few workbenches at the edges of the room. Our class was twenty lads and the teacher, who told us with relish he had been a builder all his life and had the hands to match. There were no

My first memory of the moon is looking through my brother's telescope as a boy. This photo of the moon was taken with my iPhone (looking through a telescope).

Growing up I didn't
get on at school
and channelled my
frustrations into
Sunderland Football
Club. It was the 1980s
and I got drawn into the
uglier side of sport –
fighting after games.

But astronomy was my salvation. Working as a bricklayer by day, by night I would dream of the stars. I joined the Sunderland Astronomical Society (SAS), and started to travel to star camps around the UK.

I realized the darkest and clearest skies in the UK were on my doorstep, above Kielder Forest in Northumberland. On evening trips I regularly saw thousands of stars every night, and even the aurora borealis, also known as the Northern Lights.

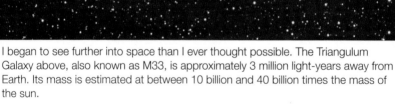

I began to see further into space than I ever thought possible. The Triangulum Galaxy above, also known as M33, is approximately 3 million light-years away from Earth. Its mass is estimated at between 10 billion and 40 billion times the mass of the sun.

Inspired by similar events around the UK, I set up a star camp in Kielder with some friends but soon we wanted something more permanent. I helped to build the Cygnus Observatory with the SAS (above) in Washington, Tyne and Wear in 2002, with bricks that I had donated from work. It was a success but we already had our sights on building something bigger . . .

The road to Kielder Observatory . . . we began fundraising and scouting for potential locations in Kielder Water and Forest Park. Encompassing over 230 square miles, the densely wooded forests burst with wildlife, home to half of England's red squirrels, as well as a multitude of deer, goats, goshawks and buzzards. But the dark skies above are really what attracted us.

Kielder Water and Forest Park is home to numerous sculptures, including James Turrell's acclaimed Skyspace (bottom image). We wanted the observatory to be a leading work of architecture, too, and launched a competition with RIBA to find a winning design.

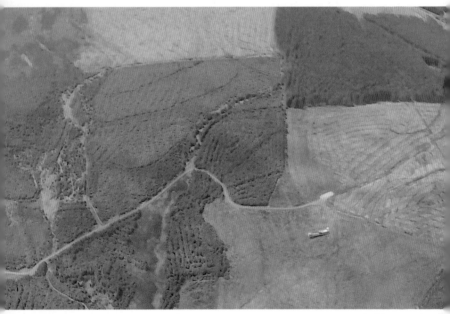

After years of fund-raising, and a closely-fought competition with entries submitted from around the world, we finally settled on a design and a location. My dream of building a safe place where anyone could see the stars was taking shape.

desks where we could sit and study; instead of pens and paper, a labourer delivered us each a pile of bricks, along with a model plan. It could be anything from a chimney stack to a brick fireplace, which were oddly popular then. Our job, from then until the end of the course, would be to build anything put in front of us.

Each morning was the same routine: arrive at the workshop and grab my weaponry from the rack on the wall. A few minutes later, spirit level in one hand, trowel in the other, I would pull on my steel-toed boots and adjust my hard hat: armed and ready for the day's battle. I felt like I was working to defend my family.

The hardest part of bricklaying was the hand–eye co-ordination, getting used to holding the trowel in a certain way to load enough mortar onto the blade to be able to spread it evenly and, more importantly, quickly. It taught me to take pride in what I did; if you step back and it looks good, it's probably been built correctly. The bricks were heavy and the mortar contained lime, which would strip the skin from your fingers, but I felt like a lad here, and it was easy to be who my peers wanted me to be. We would swear and tell jokes and stories – it was easy. Later I would realise the skill was in being able to lay as many bricks as possible as well as possible; this job was about volume – the more bricks you laid, the more you were paid.

I would mimic the teacher, badly at first, but it really was a case of the more bricks you laid, the better you became. I learned the knack and I was enjoying life at college, making friends quickly, with work as the bond. After half a year, we were ready for onsite work with qualified bricklayers. Although there was no graduation ceremony, I was now on the way to becoming a bona-fide brickie. Countless people were telling me how lucky I was, using the old adage, 'You'll never go hungry if you have a trade, Gary.' I always laughed at that. Yes, I was grateful, but I also hoped the sum total of my existence wasn't just to avoid hunger!

I was counting on the reality of life on a building site matching up with the vision I had in my head: hard men working in tough conditions, forearms like Popeye and a steely determination, preferably encased in years of heavy drinking and rowdy behaviour. Thankfully, it was almost exactly as I envisaged in the early years. Work was full of my kind of people, the people I had grown up with, and I felt like I had their respect. I still went to the football at weekends, but now that I had more of a purpose, and a new gang to spend time with on the job, I didn't feel the need to fight. I never went back to it. It helped that as the 1980s moved into the 1990s there seemed to be a boom in construction. I was always in work and I finally felt like I was providing for my family and fulfilling the role of father and husband. Gary soon had two new brothers, Graham and James, and finally a young sister, Stephanie, with her big brown eyes. There would be no more dole queues.

*

'Eh, lads. Do you know the universe doesn't have a boundary?'

'You what?'

'The universe. It doesn't even have a physical boundary.'

'Shud-up, Professor!'

Astronomy had taken a back seat in my life for a few years now, but as my routine at home and work settled, my interest in the stars gravitated back into focus. Towards the end of my shifts at the building site, I started to bring it up with the lads. I felt emboldened and we had some wonderful chats full of brain-teasers – never underestimate the curiosity of a brick-layer after a ten-hour shift.

Approaching my thirtieth birthday, I was walking over the Wearmouth Bridge after a Sunderland home game with a few mates. It was late winter and the light in the evening was fading. Off to the north-west I could see a faint orange glow in the sky. It was Arcturus, or Alpha Boötes, the brightest star in the constellation of Boötes, the Herdsman. I hadn't looked up at it since I was a teenager, after I had read about it in the library one afternoon in detention. Like muscle memory, facts started to come back. When we reached the pub, I couldn't resist waffling away to the lads about the physics. Sitting off the handle of the Plough, Arcturus was a red giant, I remembered, a dying star in the final stages of its existence. A little more massive than our own Sun, thirty-seven light years away from the Earth, I pronounced to my drunken congregation, it would soon swell up and expel its guts out into space, so that its light would disappear from our skies. At thirty-seven light years away, this may have already happened, because whenever we look at Arcturus we are looking thirty-seven years back into the past (a light year being the distance light travels in one year). There was another round of 'Shud-up, Professor!' but I laughed and knew they were impressed. I felt I could tell who I wanted about astronomy and they could no longer hurt me for it; I felt bigger.

Around this time, in 1995, I started to hit the science books again in the evenings, ranging from simple guides to the constellations to Stephen Hawking's *Brief History of Time*. It was hard fitting the reading in, considering I got home late from work and had four children to take care of with Maureen. But as I soaked my aching muscles in the bath late at night, there was nothing better to help me unwind than contemplating the depths of four-dimensional space–time, at least I thought so. The next big step in my celestial resurgence was buying a new telescope.

I trawled the phone books until I found a local dealer in Ryton, near Newcastle. I called ahead and a kind man picked up, who was generous in answering the questions of a novice. I didn't know it then, but David Sinden was a big wheel in astronomy. He was old-school, an ex-optical engineer at the world-renowned telescope-making firm Grubb Parsons, who had built some of the world's largest telescopes before the firm changed its remit and began building cheaper telescopes. Now, as his career wound down, he had set up a new shop to go back to basics and help amateur astronomers. When I arrived at David's showroom I was overwhelmed. The place was empty of customers, but telescopes were scattered on every surface: big ones, small ones, some on tripods and others lying in heaps on the floor. The room was cramped and its white walls were plastered with images of nebulae and galaxies, some I excitedly recognised. I could hear the sound of humming machinery out the back. I called out, but no one answered, so I walked behind the counter and peeked into the backroom. There he was, lab coat on, bushy grey-white beard and tools in hand. David was fitting a new mirror to a telescope. He eventually looked up and turned off his machine. I could tell instantly that he lived for astronomy, and his enthusiasm engulfed me as he manically tried to explain the complex optical problem he was hoping to fix. I didn't have much of an idea what he was talking about, but I was still delighted. For the first time in my life I had discovered another astronomer. It felt like being reunited with a long-lost relative, but one who was smarter and more profound, and with a very impressive beard.

We spoke for nearly an hour and, after parting with £300, I left with a 100mm Meade reflecting telescope under my arm, several eyepieces and more excitement than I could handle. David explained that reflecting telescopes use a mirror to focus the light from a distant object down a straight tube. The focal length is the distance light travels after reflecting off the primary mirror until it reaches focus, which in this case was 1000mm. The cost per inch of aperture – the width of the telescope and how much light is let in – is much higher in refracting telescopes than for reflecting telescopes, which are the other kind and use a glass lens to focus the light.

Once home, the kids descended on me. 'What is that, Dad?!' We set it up in the back garden and assembled it carefully, its dull metallic smell new and intoxicating. Up until now I had only used manual point-and-shoot telescopes, which are ideal for beginners, but this one had motors and an automatic tracker for the more experienced and adventurous. I inserted the batteries which would enable the telescope to track the sky against the Earth's rotation. David had explained to me the reason for this in the shop. Because our planet spins on its axis once every twenty-four hours (or twenty-three hours, fifty-six minutes and 0.49 seconds, to be precise), this regular rotational speed means that if you are looking at the night sky through a telescope, the stars will gradually appear to move across the sky (and eventually out of your line of sight in the scope). This isn't so important for observing through a telescope, but if you want to take pictures through the scope (by replacing the eyepiece with the lens of your camera), then the star or object you are taking a picture of in the sky will move as you click away, particularly if you have the camera on a long exposure. To get the best image, you need to keep the star as still as you can. To do this you need a telescopic mount that moves in exactly the opposite direction and speed of the Earth's rotation to counteract the star's movement. So once I had inserted the batteries, my new telescope's mount would take twenty-three hours, fifty-six minutes and 0.49 seconds to do one complete rotation. I marvelled at this neat trick.

The set-up was nearly complete, but my children had now lost interest and gone inside. The digital handset connected to the telescope glowed red, as did the buttons, so as not to affect my dark-adapted vision. I aligned the scope and set off trying to find something, anything at all, using the automatic digital finder. The display read 'M36'. I had never heard of this object before, but off I went, the scope automatically clicking and moving across the sky. After a few minutes it stopped, aimed at a north-east-orientated point above. In went the eyepiece; I had learned my mistake from my first time as a boy with the Moon. I started rotating the focusing wheels, focus came and slowly an image appeared. I could see a cluster of stars, just faint white dots, but undeniably there against the blackness of space. I was elated and shouted for my whole family to come out and join me.

But no one came – they were all in bed. I had been outside for so long getting everything set up that I had lost track of time. It was now 2 a.m. and there was just M36 and me. But I didn't care; I had a new telescope and astronomy was back on the agenda. I hadn't felt as excited as this for years.

I began subscribing to astronomy periodicals like *Astronomy Now* and *Sky & Telescope*. I learned that M36 is a cluster of over sixty stars in the constellation Auriga, all more than 4,000 light years away from Earth. I soon wanted to see more and more, and further and further. But I quickly realised something important: I needed a bigger aperture on my telescope if I was to travel further in space, and further back in time from my garden in Sunderland. Thankfully, David Sinden was my guide yet again. He taught me to think of a telescope as a giant light-bucket. All it is doing is collecting more light for us than our eyes can capture on their own. And just as a wider bucket collects more rain, a wider telescope collects more light. So a telescope's width or aperture is one of the most important factors to consider when making a purchase. Aperture tends to be measured in inches, and my current scope had an aperture of 4 inches.

I had for some time been looking at a 10-inch-aperture telescope called a Meade LX200, but it was way out of my price range. I would stare at it in my magazines and on websites – I had aperture fever – until one Sunday afternoon when I was introduced to Graham Darke, another stargazer who would become one of my most trusted friends and mentors, affectionately known to me as 'Dickie'.

Yep, here was an amateur astronomer whose surname happened to be Darke, but even more conveniently, he lived next door. It was the winter of 1996 and the Sunday morning my son Graham had burst into our house, shouting that the bloke next-door had laid his hands on my 'dream machine'. Graham had been in Dickie's kitchen, collecting his fee for washing Dickie's car when he recognised the object I'd been pining over for months. I couldn't believe it when he told me at first, but as soon as I went over to investigate that afternoon, there it was, bulky but magnificent. Classified as a Schmidt-Cassegrain, a reference to the surnames of the two engineers who invented its optical system, its short stubby tube was painted a deep metallic blue, set off by the jet-black anodised fork cradle which held it firmly to the mount. A large metal lid protected the optics, and the cables and battery were neatly fitted to the side. I was drooling.

Dickie could sense my eagerness and that evening agreed to pop over for an impromptu observing session in my back garden. I waited excitedly until around 9 p.m., when through the curtains, I watched him come over. Over six feet tall and thickset, he didn't seem to have much trouble carrying the telescope under his arm. He was friendly and unassuming as we set up our two scopes next to each other, humbly describing his day job as an insurance salesman. I sensed that, like myself, he too lived for astronomy in the evenings. But that was where the comparison ended – he was a true pro. I thought I knew my stuff: the theory, my way around the constellations and the basics of the telescope. Dickie was in a different league; if the intricacies of the night were a craft, he was a master.

I watched silently as he calmly took out his tripod, assessing for a few minutes the optimum height the legs should be adjusted to, both for a good view and to save his back from bending over. I'd never given it that much thought before, but casually mimicked him as if I did this every time as well. Although I liked Dickie instantly, and felt comfortable with him, I didn't want to be seen as a weak lad asking for help, at least not just yet.

Once the tripods were erected and secure, we attached our telescopes. I couldn't help staring at his LX200, which dwarfed my lowly 100mm, and did a quick mental calculation – his telescope gathered more than six times the light mine did (π r2).

'What magnification can you get with the LX200?' I asked hastily. I couldn't wait to look through it and had a few targets in mind.

'Well, telescopes don't actually magnify at all,' Dickie answered patiently.

I must have looked baffled, so he kindly continued.

'Basically, the shape of the telescope's mirror or lens gathers the light and forces it to converge at a point. Imagine looking down an ice-cream cone: the wide end of the cone being the front of the telescope and the thin end of the cone being where the eyepiece is located. We call the point at the narrow end of the cone the focus, which the eyepiece then magnifies. You can calculate the magnification the eyepiece gives you by doing a little maths.'

It was starting to click and so I studied the number on my eyepiece, which was 25mm. Dickie explained that this was the focal length of the eyepiece: the distance light travels through the eyepiece before reaching focus. The telescope itself also has a focal length: the distance in millimetres light will travel before it reaches focus at the eyepiece. If you divide the focal length of the telescope by the focal length of the eyepiece, that number is the magnification value. And the higher the overall number, the higher the magnification, so that smaller eyepieces (25mm rather than 100mm) provide greater magnification.

As the evening got darker, Dickie generously continued to educate me about optics and how to navigate the sky. As with David Sinden, it felt so comforting to be talking to someone who shared my passion, and we laughed and kept the conversation going easily. I had an idea: I wanted to observe Jupiter. I had seen many images of this magical planet, a world of clouds over 350 million miles from our Sun, and so different from Earth – a ball of mostly hydrogen gas, with not even a surface to walk upon. In particular, I was desperate to see Jupiter's red spot, the largest storm in our solar system, which has been raging for at least 186 years. It can be observed as a distinctive swirling pattern of gas in the planet's cloud tops, known as the 'great red spot'.

I had no idea what to expect observing it through a telescope, but I knew it should at least be visible in the east portion of the night sky this time of year, as I had checked on my star chart. I asked Graham if we could see it and he replied with a casual 'yes' and a nod of the head, but I sensed he was keen to see it too, his eyes scanning the sky continually. Now his eyes dropped to the LX200 and he held the computerised controller to type in the instructions to find the planet. He pressed 'Go to' and, as he did so, the motors started to turn and a whirring broke the silence. 'Having a coffee, are we?' I joked, the telescope's noise surprisingly similar to a coffee grinder. The scope eventually came to rest as my anticipation peaked. What would I see? I had forgotten all about my own telescope and it was all eyes on the LX200. Graham lowered his head once more to the eyepiece. Silence again. I wanted to push him away so I could see it for myself, but I resisted the urge.

'Er, yep, it's there,' said Graham, trying to be as under-stated as he could. But after thirty seconds he gave up on the charade. 'WOW!' He smiled in my direction. 'Want to have a look?'

I nodded nonchalantly, but I wasn't fooling Dickie. Internally I was doing hoops and it showed on my face. I lowered my head to the eyepiece and there it was. I saw a ball of white light, smudgy and small at first. I focused the telescope and slowly the blur lifted. The line of Jupiter's circumference sharpened, and two distinctive bands around the equatorial plane appeared. It was very bright and gradually, as I instinctively moved my head around, observing through the wide-field eyepiece, I could see more detail: the bands appeared to have patterns in them. It was hard to see initially, but I was transfixed as more and more subtle detail began to emerge, with lighter and darker regions coming into view. Eventually, there it was: the eye of the storm – about 12,400 miles long and 7,500 miles wide, nearly three times the size of Earth. I must have looked at the great red spot for seven or eight minutes silently, gently adjusting the focus in an attempt to pull out more detail. I heard Dickie say, 'Can you see the four star-like objects?'

Star-like objects? What did he mean? I could see four dots that, yes, looked like stars, but what else could they be?

'They are Jupiter's four main moons.'

Just when I thought the night couldn't get any better, I had seen Io, Ganymede, Europa and Callisto. I had read about their discovery by Galileo Galilei in 1610. Although Jupiter has at least sixty-three known moons, these four are by far the largest. Ganymede is indeed the largest moon in the solar system, greater in diameter than Mercury and two thirds the size of Mars. I couldn't see much detail in them, such as the active volcanoes of Io which cover the moon in sulphur, or Callisto's cratered icy surface, but I could see close to what Galileo would have seen hundreds of years before, viewed with his own state-of-the-art equipment (I like to think of him with his own antique LX200). His discoveries about Jupiter landed him in hot water; they led to his blasphemous hypothesis that the Earth could not be the centre of rotation in the universe if these moons were orbiting around Jupiter. Galileo was eventually put before an inquisition and placed under house arrest for the rest of his life until his death in 1642. That night in my back garden, I thought of him imprisoned for looking at this planet and its moons. I thought, too, of everything I had been through until that point in my own life. I felt so lucky.

March–April Sky Guide

Gemini–Ursa Major–Ursa Minor–Draco–Lunar Eclipse

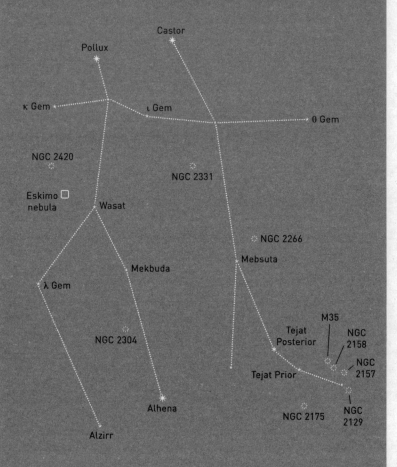

GEMINI

In spring (March) looking west from Kielder Observatory

Castor

Pollux

κ Gem

ι Gem

θ Gem

NGC 2420

NGC 2331

Eskimo
nebula

Wasat

NGC 2266

Mebsuta

Mekbuda

λ Gem

M35

NGC
2158

Tejat
Posterior

NGC 2304

NGC
2157

Tejat Prior

Alhena

NGC 2175

NGC
2129

Alzirr

Stars   Mag 0 ✸   Mag 1 ✳   Mag 2 ✶   Mag 3 ✲   Mag 4 ·   Mag 5 ·   Clusters ⬡   Nebulas ▢

Gemini is one of the famous zodiacal constellations, dominated by two bright stars known as Castor and Pollux, named after two mythical Greek twins who were the brothers of Helen of Troy, and who would later join Jason and the Argonauts on their quest for the Golden Fleece.

At 9 p.m. on 1 April these two stars can be found in the south-western sky above the dominant constellation of Orion. Whilst the Hunter Orion is beginning to head for a summer hibernation, you should still be able to see him close to the horizon. If you then join the two bright stars Betelgeuse and Rigel together, and extend that line upwards, they will lead you directly to the white, magnitude-1.9 Castor. His slightly brighter, yellow-tinged brother, Pollux, is just to the left. The fact that no two other bright stars appear this close together in the sky might explain why the ancients assigned the myth of the twins to them. It is also worth pointing out that Castor is often given the designation Alpha Geminorum (α Gem) despite not being as bright as its twin.

Castor itself is not a single star, but instead a staggering six stars. Even under relatively low magnification, you will see Castor split into two component stars. A third star close by is also gravitationally bound to this pair. But here's the kick: each of those three stars is also

91

a double star. These additional companions aren't visible to amateur instruments, but professional astronomers have teased out their presence by analysing the light coming from the region. They have also been able to confirm that Pollux is orbited by a planet known as Pollux b, which is around 2.3 times heavier than Jupiter.

Gemini is not the richest of constellations for deep-sky objects, but there are a few noteworthy targets. Perhaps the easiest to see is the open cluster M35. With a magnitude of 5.3, it should be visible with the unaided eye to those with darker skies. City-dwellers will need to get out a pair of binoculars. The cluster contains about 200 stars and is located approximately 2,800 light years away. Gemini contains several other star clusters which are accessible with binoculars, including NGC 2129 and NGC 2355.

For those with a medium- to large-aperture telescope, you could point it at the magnitude-10 Eskimo Nebula (NGC 2392). It gets its name from the fact that the expanding shell of gas resembles a parka hood pulled over someone's face. Discovered in 1787 by William Herschel, six years after he made his name by discovering the planet Uranus, it is what astronomers refer to as a 'planetary nebula' – the death throes of a star like the Sun.

The star that formed the Eskimo Nebula eventually ran out of hydrogen and for a time started to convert helium into carbon. This process created a lot more energy than before and so the star bloated outwards, probably devouring any planets that once orbited it. Helium-burning is very unstable, however, and so the star eventually shook itself apart and we see the result as a beautiful nebula. Our Sun will succumb to a similar fate in around 5,000 million years.

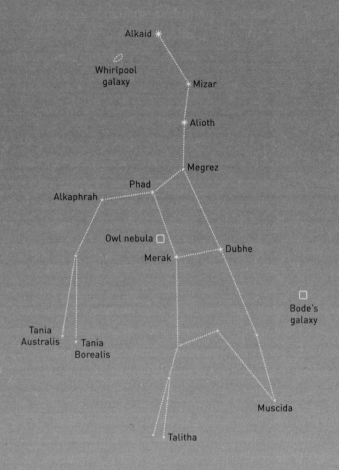

URSA MAJOR

In spring (April) looking north from Kielder Observatory

Alkaid

Whirlpool
galaxy

Mizar

Alioth

Megrez

Phad

Alkaphrah

Owl nebula

Merak

Dubhe

Bode's
galaxy

Tania
Australis

Tania
Borealis

Muscida

Talitha

Stars   Mag 0 ✸   Mag 1 ✷   Mag 2 ✴   Mag 3 ✦   Mag 4 ·   Mag 5 ·   Clusters ⬡   Nebulas ☐

Along with Orion, Ursa Major is one of the most recognisable constellations in the sky, and is also called the Great Bear. For reasons outlined in Chapter 1 of this book, particularly famous are the seven stars in the shape of a giant saucepan, known by various names around the world, including the Plough and the Big Dipper. However, it is wrong to call these seven stars alone the Great Bear as they only make up her rump and tail. Instead, astronomers have another name for familiar patterns either within or across constellations: asterisms.

For many observers in the northern hemisphere Ursa Major is circumpolar – it never sets below the horizon and so is visible all year round. March and April, however, is a particularly good time for observing it as it sits high up in the sky. The Plough asterism is really hard to miss, although people new to the stars sometimes don't realise just how much of the sky it covers (Ursa Major is the third largest constellation). So be sure to cast your eyes wide over the north-eastern part of the sky. I recall a group of schoolchildren visiting us from Sunderland and they pointed out to me that the asterism looked like a shopping trolley.

The four stars that make up the 'bucket' of the saucepan are the easiest to find first – Dubhe, Merak,

Phad and Megrez. From them, you can trace down the three stars which form the 'handle' of the pan – Alioth, Mizar and Alkaid. Even without binoculars you should notice that Mizar is a double star, with a companion known as Alcor. In fact, Mizar is a quadruple star system and Alcor by itself is a binary system, meaning there are six stars here in total.

Unlike Gemini, Ursa Major is bursting with deep-sky objects to explore. Let's start with a wonderful pair of galaxies – M81 and M82. To find them with binoculars or a telescope, extend an imaginary line between Phad and Dubhe on opposite corners of the bucket of the Plough towards the north. You're aiming for the magnitude-4.5 star 24 Ursae Majoris. From here you want to head back eastwards towards the magnitude-8 star HIP 49230. Before you reach it, however, and after a journey of about 6 degrees, you should encounter the spiral galaxy M81 (also known as Bode's Galaxy). With binoculars and small telescopes it will appear as a fuzzy blob, but with apertures above 8 inches you should start to make up some of its spiral structure. It is particularly well positioned for observation as we are seeing the galaxy face-on.

Under low magnification you'll also be able to get the nearby M82 in the same field of view, though you won't see it in as much detail. It is known as a 'starburst galaxy' because it is undergoing a particularly high rate of star formation. On 21 January 2014, a supernova was discovered in the galaxy by a team at University College London. Reaching a peak brightness of magnitude 10.5, it was visible to amateurs too.

As if that wasn't enough, there are more Messier-listed galaxies to search for in Ursa Major. The first is the famous Pinwheel Galaxy (M101), located further down towards the Bear's tail. Take Alkaid, the star at the tail's tip, and head towards the north and the star HIP 68304. M101 is about 2.5 degrees further on in the same direction. It was made famous by the stunning image taken with the Hubble Space Telescope, but be warned that its low surface brightness can make it a tricky and often frustrating target for amateur astronomers.

Moving into the body of the Bear, not far from Merak, you'll find M108. If you slide less than a degree towards Phad, you'll see the star HIP 54314 shining at magnitude 7.25. Below it are four stars in the shape of a curve and following them will lead you straight to M108. Then, less than half a degree away, there is another planetary nebula to look at – the Owl Nebula (M97). Like the Eskimo Nebula in Gemini, it is the remnant of a small star like the Sun.

Our final stop is the galaxy M109. Head down the base of the Plough towards Phad and overshoot it by around half a degree and the galaxy should be in your sights. It is a barred spiral galaxy, and we are looking at it face-on.

(It is worth a quick note to say that the famous photograph known as the Hubble Deep Field was taken in this region, close to Megrez. It revealed thousands of galaxies occupying an area of the sky just 2.5 arcminutes across.)

URSA MINOR

In spring (April) looking north from Kielder Observatory

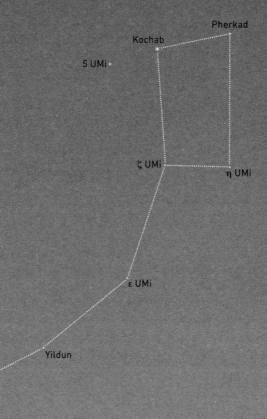

Pherkad

Kochab

5 UMi

ζ UMi

η UMi

ε UMi

Yildun

Polaris

NGC 188

Stars   Mag 0 ✳   Mag 1 ✷   Mag 2 ✶   Mag 3 ✦   Mag 4 •   Mag 5 ·   Clusters ⁘   Nebulas ▢

The stars of Ursa Minor, or the Little Bear, form a similar shape to those in the Plough, leading to its other nickname of the Little Dipper. Many of these stars will be washed out in highly light-polluted areas and are very difficult to see. That said, you will certainly be able to make out the brightest star in the constellation: Polaris. As I outlined in the Preface and in Chapter 1, it is also known as Polaris, or the North Star, and is arguably the most important star in the night sky.

As the Earth rotates, the stars seem to be moving across the sky over the course of an evening. But because Polaris sits almost directly above our axis of rotation, it barely seems to us to move at all.

To recap (see Chapter 1 for more detail), locating Polaris for yourself is straightforward. Start with the seven stars of the Plough in Ursa Major. Then take Dubhe and Merak, the two stars furthest from the handle. If you join the line between them and extend it upwards from Dubhe, you will be carried straight to Polaris. For this reason Dubhe and Merak are often known as the Pointers.

As a relatively small constellation, Ursa Minor doesn't have many deep-sky objects of note. Perhaps your best bet is the barred spiral galaxy NGC 6217. With a magnitude of 11.2, you'll need at least a 4-inch aperture telescope to pick it out. It sits about 2 degrees below the line between ζ UMi and ε UMi and is nestled at the heart of a rectangle formed by the four stars HIP 80480, HIP 80850, HIP 81854 and HIP 81428. Like M82 in Ursa Major, it is a starburst galaxy.

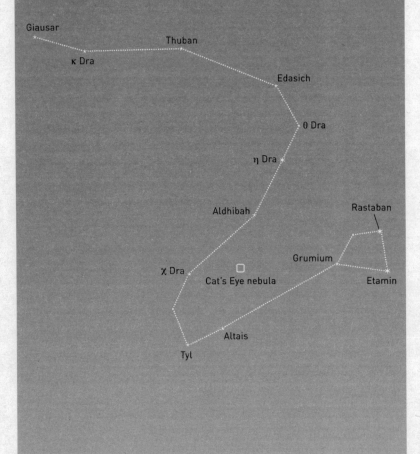

DRACO

In spring (April) looking east from Kielder Observatory

Giausar

κ Dra

Thuban

Edasich

θ Dra

η Dra

Aldhibah

Rastaban

Grumium

χ Dra

Cat's Eye nebula

Etamin

Altais

Tyl

Stars   Mag 0 ✹   Mag 1 ✴   Mag 2 ✳   Mag 3 ✦   Mag 4 ·   Mag 5 ·   Clusters ⁑   Nebulas ▢

Draco is a long and winding constellation that seems to snake across the sky, hence its name, which comes from the Latin word *draconem*, which means 'large serpent'.

If you've traced the line between Dubhe and Merak in Ursa Major up to Polaris, you will have skimmed by the star Giausar (λ Dra) at the tip of the dragon's tail. That tail and then his body stretch downwards and eastwards, running parallel to the handle of the Plough, before bending back westwards and underneath Ursa Minor. Just after it changes direction you'll encounter the magnitude-2.7 double star η Dra. Continuing onwards you eventually reach the four stars which make up his head – Grumium (ξ Dra), the double star Kuma (ν Dra), Rastaban (β Dra) and Etamin (γ Dra).

The brightest star in Draco is Thuban, which can be found near the dragon's tail approximately halfway between Mizar in Ursa Major and Kochab in Ursa Major (β UMi). Around 5,000 years ago it was Thuban, not Polaris, that was our Pole Star. This is due to an effect known as precession, which sees the Earth's axis wobble as our planet is jostled by the gravitational pull of the Sun and the Moon. As a result, the North Pole traces out a circle which takes 26,000 years to complete. So eventually the North Pole will point away from Polaris, and Thuban will return to being our Pole Star in around 21000 CE. The entire constellation is circumpolar for most northern hemisphere observers.

The highlight of the deep sky in Draco is definitely the Cat's Eye Nebula – the subject of one of the Hubble Space Telescope's most stunning images. Like the Eskimo Nebula in Gemini and the Owl Nebula in Ursa Major, it is a planetary nebula. To locate it, draw an imaginary line between the stars δ Dra and ζ Dra. Around the midpoint of this line sit two more stars: the magnitude-6.85 HIP 88583 and magnitude-7.65 HIP 87530. The Cat's Eye Nebula lies almost exactly between them.

It was discovered by William Herschel in 1786 and he was the one to name these objects 'planetary nebulae' because to him they looked like planets through his telescope. However, when the Cat's Eye became the first planetary nebula to have its spectrum studied in 1864, it became clear that it was gaseous in nature.

Under low magnification it will look like a fuzzy blue-green blob. If you can ramp up the magnification, then you may start to see some more of its structure. More advanced observers may want to add an OIII (Oxygen 3) filter to their telescope, which will increase the contrast of the nebula against the background sky.

Sliding down Draco to where his tail meets his body, we find the star Edasich and enter the realm of several galaxies. Within 1.5 degrees down and to the left sit three all huddled together: the 'Draco Trio' of NGC 5981, NGC 5982 and NGC 5985. They are nestled so close to one another that you should have no problem getting all three in the same field of view. The most visually pleasing is arguably the lowest of the three, NGC 5985 – a face-on eleventh magnitude spiral galaxy. NGC 5982 shines with a similar magnitude but is a more featureless face-on elliptical. At magnitude 13, NGC 5981 is the hardest to make out because we are seeing it edge-on.

Skipping back across Edasich and heading in the direction of HIP 73837 will take you to another galaxy – NGC 5866. Many astronomers believe this to be the missing M102. In Charles Messier's list of deep-sky objects, it isn't clear which galaxy he labelled M102. However, NGC 5866 is a good candidate as it matches the position of the object scrawled in his handwritten notes. It is also a good fit with the description given of M102 by Pierre Méchain when the list was published in 1781. It is known as a lenticular galaxy, meaning it is part way between an elliptical galaxy and a spiral galaxy.

Lunar Eclipse

Eclipses are some of the most spectacular sights in astronomy, occurring when the Sun, Earth and Moon all line up. Astronomers have a vowel-less word for three celestial objects in a row: a syzygy. This does not happen every month, because the Moon's orbit is inclined to the Sun–Earth line by around 5 degrees; so most months it either dips below the line or rises above it. However, sometimes the Moon passes directly behind the Earth and moves into our planet's shadow, causing a lunar eclipse.

You might imagine that no light from the Sun can get to the Moon at all here because the Earth is in the way. Yet a lunar eclipse is famously red in colour and is often nicknamed 'the blood Moon'. As the Moon is incapable of making its own light, the Sun is the only possible source of this illumination. The layers of gas in Earth's atmosphere bend some of the Sun's light around our planet to reach the Moon. If a lunar eclipse occurs just after a volcanic eruption, the ash can contribute to the bending and the Moon appears even redder.

There are actually three types of lunar eclipse, each based on which part of the Earth's shadow the Moon sits in. As the Sun is wide, and not just a single point of light, the Earth's shadow comes in two parts: the main, dark shadow directly behind the Earth (called the umbra) and a lighter, fuzzy outer shadow (called the penumbra). If the Moon is completely consumed within the umbra, we see a total lunar eclipse. If instead the Moon is straddling the border between the umbra and penumbra, we see a partial lunar eclipse, with only some of the Full Moon obscured by the Earth's main shadow. Finally, a penumbral lunar eclipse occurs when the Moon only passes through the fuzzy outer shadow. This is the least noticeable and least spectacular of the three.

The next three total lunar eclipses are scheduled for 31 January 2018, 27 July 2018 and 21 January 2019. Such events have been recorded by humans for centuries. One of the most famous lunar eclipses occurred on 1 March 1504, when legendary explorer Christopher Columbus was holed up in Jamaica. At first the locals were very hospitable and welcomed his crew with open arms. But they grew increasingly weary of Columbus's men stealing from them, and so after six months they snapped and decided to stop feeding them.

Looking for the best way out of the situation, Columbus remembered that he had an astronomical almanac on board. It contained the prediction of a lunar eclipse on 1 March. Thinking quickly, he told the locals that he was in touch with God and that He was displeased with the islanders' attitude. To demonstrate his wrath, He would turn the Moon blood-red that night. At first the locals scoffed, but after the lunar eclipse they gave Columbus all the food he needed. Sometimes it pays to know about astronomy.

3

Aurora

My dad and I load the final cases into the boot of his Ford Sierra just as the last of the evening sun starts to fade. It is warm and we can hear the birds singing from the wood close by as we delicately arrange the back seat: thin sheets and pillows to make comfy car beds for the boys.

The flight to Malaga is early so we opt to drive through the night down to East Midlands airport. It's a special family holiday, our first trip to Spain and my mam and dad are coming along for the experience too. Maureen gets in the back with my mam, and the boys slam the doors loudly as they clamber inside, their hands grasping at the headrests, white knuckles pulling excitedly at the metal supports. 'Dad says we're going where it's sunny,' they chirp to my mam. They're beaming and bouncing around before the engine starts, but once we set off it only takes a few minutes for them both to drop off. When I look back in the rear-view mirror, their foreheads reveal little beads of sweat clinging to their blond hair.

The roads are quiet and my dad and I chat softly. It is summer 1990 and England is playing in the World Cup finals in Italy. We weigh up the chances of going all the way, but then think better of it. 'Dad, do you still have your *Star Trek* tapes?' Much safer ground. We laugh with the ladies and try to imagine the warm sandy beaches with gin and tonics waiting for us on arrival at the Costa del Sol. After a little while, with my eyelids closing, we pull into a lay-by on the A66 for a break and to stretch our legs. It is darker now and the wind has dropped. Pine trees insulate the electric glow of a nearby town and the lay-by is deserted apart from a few lorries parked up for the night. My dad wanders off to have a stroll; I stay behind. When he returns, I am silent and looking up, lost in thought. I gesture to the sky, as if I might disturb what is going on above our heads. He follows my gaze. At first they appear to be clouds – perhaps they are – but I can see stars shining through them. Gradually subtle shades of green and red grow stronger and pirouette powerfully across the night sky in fits and bursts. They are dancing, putting on a performance just for us. 'My God,' is all my dad can say as he calls for my mam to come and see. I'm transfixed; it's pulsating directly overhead, gently swaying to a different wind.

My dad has no idea what we are looking at: the ethereal curtains of colour something straight out of an alien planet that Captain Kirk could have landed on. 'We can see the lights because we're far enough north,' I explain to him. 'And the Sun is active right now.' He listens attentively and then looks over in my direction. Through the darkness I can just make out a few shadows spreading across his face, filling the familiar creases of his smile – pride. We stand together, bathing in the light.

*

I lost my dad just over ten years after that night, on his sixtieth birthday. He knew he was dying, but he didn't want to talk to any of us about it; I wish he had. My dad was a strong silent type, hardened by a painful childhood, fostered and then adopted, passed between good neighbours and anyone who would look after him. With his own family he excelled, though, providing us with a warm and caring home.

The end was so quick, I was numb. One moment we were together, the next he had leukaemia; he lost his hair and disappeared. On his last night I was in his room with my brother Anth. He asked if we would fetch him an orange lolly, of all things, so we did. I unwrapped it and held it to his cracked lips. He savoured the icy orange flavour, and I could tell he really enjoyed it. When he finished, he was silent and a tear rolled down his cheek. I think that right then he knew his time was up, but still he wouldn't talk to us. The next day I held him in my arms as he died. I knew he felt my presence there with him, as I still feel his presence with me now. He was called Alan Thomas Fildes and I miss him every single day.

I cherish the summer night we had standing in the lay-by under the colours: the first time my dad and I saw the Northern Lights, or the aurora borealis. At that point in my life, I think my dad was proud of what I was doing, but he didn't think astronomy would amount to anything more for me than a hobby after work. The sadness is that it would be different now. I didn't begin to get really serious about it until 2000, the same year he died. I wish I could have spent more nights under the stars with him, chatting about the things I know now, particularly about the aurora.

I've always been taken with the myths that surround them, of how mankind made sense of these unearthly lights hundreds and even thousands of years ago. Ernest Hawkes, an American anthropologist who studied indigenous peoples around the Arctic Circle, recorded some of the most magical beliefs. For the Labrador Inuit, the lights were the torches of departed souls in heaven, guiding the way for new arrivals to the afterlife like an ancient runway landing strip. For the Inuit of Greenland and Alaska, the origins were stranger still. Spirits could be observed celebrating within the colourful lights of the aurora itself, and even playing football with a walrus skull. Knud Rasmussen, a Danish explorer, recorded that 'it is this ball game of the departed souls that appears as the aurora borealis, and is heard as a whistling, rustling, crackling sound. The noise is made by the souls as they run across the frost-hardened snow of the heavens. If one happens to be out alone at night when the aurora borealis is visible, and hears this whistling sound, one has only to whistle in return and the light will come nearer, out of curiosity.' Scientifically, the aurora's accompanying noise has never been proved or explained, although many people have claimed to hear it to this day. Some theorise that the brain may detect electromagnetic waves from the lights and, through some process unknown to us, convert them to sound. Another theory suggests that the aurora creates electrical discharges on Earth from objects like buildings, wire fences and trees, which the human ear can hear. I've never heard them myself, but they add to the mystery of the Northern Lights. I've never listened out for the voice of my dad either, or whispered back, but his spirit is never too far away when I see the aurora, or when I gaze up into the sky for that matter.

Not long after I returned from that holiday in Malaga in 1990, I became intent on seeking out the aurora again for myself. The lights are notoriously elusive, which only makes their appeal more alluring, but understanding how they work reveals what makes them truly remarkable, particularly when you consider the astonishing processes required for them to even occur. I would have to find dark, clear skies to see them, but first I had to travel to the centre of our solar system, to understand where the aurora was truly born.

*

Fifteen million degrees Celsius and crushing gravitational pressure. These are the conditions required at the heart of our Sun to generate the energy to seed the aurora borealis, a term first coined by Galileo, who combined the name of the Roman goddess of dawn, Aurora, and the Greek name for the North Wind, Boreas. The aurora's odyssey starts deep in the Sun's core, in a process known as nuclear fusion, whereby hydrogen atoms, the lightest and most abundant element in the universe, are forced together at immense pressure to produce the element helium. Four hydrogen nuclei are fused together to make one helium nuclei, and the mass that is lost in the process is converted into heat and light energy. A particle's journey to the Sun's outer limits is long and hazardous as it endures many collisions per second with the Sun's atomic structure. The electrically charged medium through which they flow is known as a plasma, and because of the Sun's vast size and the high frequency of the collisions, it can take many thousands of years before the particles even reach the Sun's cooler exterior. If you have waited years to see the aurora, it's worth considering how long a single display has really been in the making.

Once at the surface of the Sun, the plasma is held there gravitationally, but the high-energy environment can blow millions of tonnes of mass per second into space, either scattered in a steady stream of particles in all directions, known as a solar wind, or becoming trapped in the Sun's enormous magnetic field. In the same way you pull an elastic band tighter and tighter until eventually it snaps, the Sun's magnetic field traps more and more energy until eventually it becomes too much and the magnetic field twists and breaks. When this happens, an enormous eruption of hot plasma is released into space, known as a solar flare, which adds energy to the solar winds. The biggest of these solar flares, a coronal mass ejection (CME), can rocket out a billion tonnes of material at speeds in excess of a million miles per hour. These flares can be packed with enough energy to power the continents on Earth for a million years, and if they are pointing towards us, aurorae will normally reach our skies within one or two days.

For aurora chasers like me, this is where the fun begins. The solar flare is the starting gun; the two-day countdown is under way. For the next forty-eight hours I track the aurora's progress like a trader watching his stock portfolio. I begin in front of a computer screen, staring at mounds of data. Websites like SpaceWeatherLive.com give a real-time update of the aurora forecast. They rely mainly on data from satellites such as NASA's ACE spacecraft, which stands for 'Advanced Composition Explorer', and the SDO, the Solar Dynamics Observatory. Both have a prime view of the travelling solar wind from its orbit in space between the Sun and the Earth. Launched in 1997, ACE orbits at a position of gravitational equilibrium, which keeps it held in position about 1.5 million kilometres from Earth and 148.5 million kilometres from the Sun. As well as offering advanced warning, accurate to within an hour, about how aurora activity may look on Earth, it identifies high-energy particles travelling towards the Earth that could potentially overload power grids, disturb communication satellites or even present a danger to astronauts on board the International Space Station.

When it approaches Earth, the first thing that the solar wind will interact with is the Earth's far-reaching magnetosphere, which points north. As a result of the nuclear decay of elements deep in the Earth's core, we retain a molten centre of rock. As this core rotates, frictional forces induce a magnetic field that propagates off into space, protecting our planet from harmful deep space and solar radiation. For the most part, our 'force field' does an excellent job at deflecting solar winds accelerated by flares. However, there are chinks in our armour. When I am aurora-hunting, the most important measurement I am checking for is the Bz orientation.

The Bz is a component of the Earth and Sun's magnetic fields, known as the IMF (interplanetary magnetic field), propagated by disturbances in the solar wind. If the orientation of the Bz is south, the magnetic field of the solar wind may cancel out the Earth's magnetic field and essentially create a zip wire for particles to cascade into our atmosphere.

Once this happens, highly energetic particles arriving from the Sun in turn set up electric currents in our atmosphere near the North and South Poles. These tiny particles interact with oxygen atoms and energise the electrons, which jump to higher orbital levels. In this state, at around 150km above Earth, the electrons want to return to their original orbits, so they decay, and in the process of doing this they release a photon of light, which can appear red or green. As the particles fall further in orbit, they will interact with nitrogen lower down in our atmosphere, and from there they produce blues and crimsons at around 80km above Earth. These colours are what we see as the Northern or the Southern Lights, and if you are lucky, you can observe all of these shades in an auroral pillar, which I like to call the stacking effect.

During a particularly big CME, the Earth's magnetic field can become so overwhelmed that northern auroral activity is pushed far beyond its normal home around the Arctic Circle. Many times a year the lights are visible in the north of England – from the dark skies in Kielder, for example. The fiercest CME on record was the so-called

Carrington event of 1859, when sailors reported seeing the aurora as far south as the Caribbean.

Having said that, the further north you are in the northern hemisphere or south in the southern hemisphere, the more likely you are to be able to catch them. Knowing what time of year to go is a notoriously tricky question. The aurora is unpredictable as it is mainly based on the magnetic mood swings of the Sun. However, there are certain things we can say. Generally in my experience it is best to go between September and April, as during the other half of the year it remains too bright at night. Colder nights sometimes provide more of a chance, when there is less humidity in the atmosphere to obscure the lights. The Kp index (the global geomagnetic storm index) is one of the most useful ways to predict how the aurora will appear in your part of the sky. Drawing on a series of indicators gathered from the Earth's space environment, including geomagnetic activity, the scale ranges from a Kp reading of 0 to 9: the lower the figure, the more northerly the location you need to find yourself in to see the aurora. On a night when the Kp index is 6, you would have a good chance of seeing the lights from somewhere like a dark-sky site at Kielder in Northumberland, but if the Kp is a 2, you would need to be somewhere further north such as Norway (see table below). This is a good estimating system; however, it does only provide an estimate. Many factors will need to fall into place to see the aurora, such as a south-pointing Bz. I have been in Kielder with a Kp of 6 and yet have seen nothing; but I have also had a Kp of 3 with a south-pointing Bz and boom, it fills the sky.

Kp	Visible from	
0	North America:	Barrow (Alaska, USA) Yellowknife (Northwest Territories, Canada) Gillam (Manitoba, Canada) Nuuk (Greenland)
	Europe:	Reykjavik (Iceland) Tromsø (Norway) Inari (Finland) Kirkenes (Norway) Murmansk (Russia)
1	North America:	Fairbanks (Alaska, USA) Whitehorse (Yukon, Canada)
	Europe:	Mo i Rana (Norway) Jokkmokk (Sweden) Rovaniemi (Finland)
2	North America:	Anchorage (Alaska, USA) Edmonton (Alberta, Canada) Saskatoon (Saskatchewan, Canada) Winnipeg (Manitoba, Canada)
	Europe:	Tórshavn (Faroe Islands) Trondheim (Norway) Umeå (Sweden) Kokkola (Finland) Arkhangelsk (Russia)
3	North America:	Calgary (Alberta, Canada) Thunder Bay (Ontario, Canada)
	Europe:	Ålesund (Norway) Sundsvall (Sweden) Jyväskylä (Finland)

Kp	Visible from	
4	**North America:**	Vancouver (British Columbia, Canada) St John's (Newfoundland and Labrador, Canada) Billings (Montana, USA) Bismarck (North Dakota, USA) Minneapolis (Minnesota, USA)
	Europe:	Oslo (Norway) Stockholm (Sweden) Helsinki (Finland) St Petersburg (Russia)
5	**North America:**	Seattle (Washington, USA) Chicago (Illinois, USA) Toronto (Ontario, Canada) Halifax (Nova Scotia, Canada)
	Europe:	Edinburgh (Scotland) Gothenburg (Sweden) Riga (Latvia)
	Southern Hemisphere:	Hobart (Australia) Invercargill (New Zealand)
6	**North America:**	Portland (Oregon, USA) Boise (Idaho, USA) Casper (Wyoming, USA) Lincoln (Nebraska, USA) Indianapolis (Indiana, USA) Columbus (Ohio, USA) New York City (New York, USA)
	Europe:	Dublin (Ireland) Manchester (England) Hamburg (Germany) Gdańsk (Poland) Vilnius (Lithuania) Moscow (Russia)
	Southern Hemisphere:	Devonport (Australia) Christchurch (New Zealand)

(Continued)

Kp	Visible from	
7	**North America:**	Salt Lake City (Utah, USA) Denver (Colorado, USA) Nashville (Tennessee, USA) Richmond (Vermont, USA)
	Europe:	London (England) Brussels (Belgium) Cologne (Germany) Dresden (Germany) Warsaw (Poland)
	Southern Hemisphere:	Melbourne (Australia) Wellington (New Zealand)
8	**North America:**	San Francisco (California, USA) Las Vegas (Nevada, USA) Albuquerque (New Mexico, USA) Dallas (Texas, USA) Jackson (Mississippi, USA) Atlanta (Georgia, USA)
	Europe:	Paris (France) Munich (Germany) Vienna (Austria) Bratislava (Slovakia) Kiev (Ukraine)
	Asia:	Astana (Kazakhstan) Novosibirsk (Russia)
	Southern Hemisphere:	Perth (Australia) Sydney (Australia) Auckland (New Zealand)
9	**North America:**	Monterrey (Mexico) Miami (Florida, USA)
	Europe:	Madrid (Spain) Marseille (France) Rome (Italy) Bucharest (Romania)
	Asia:	Ulan Bator (Mongolia)
	Southern Hemisphere:	Alice Springs (Australia) Brisbane (Australia) Ushuaia (Argentina) Cape Town (South Africa)

*

Back in 1997 in Sunderland, once I had started to grasp the science, I went hunting for the aurora in the north-east of England. My neighbour Dickie and I were spending increasing amounts of time together and had forged a close bond in our nocturnal lives. All I had to do was be me – no frills and no pretending. One weekend the following year, in March, Dickie phoned me up.

'Should be a clear forecast tonight. A few of us are going down Derwent reservoir's dark-sky site. You in?'

I had been waiting for this invite. I had never been to a dark-sky site before, but I had heard so much about them: remote locations in the countryside where light pollution is minimal and the stars are especially bright. It could mean an excellent shot at the aurora. I hastily accepted and asked what I would need.

'Just bring your scope and a flask, get wrapped up.'

I placed all the bits on my bed and was ready in minutes. It was like running out the door past my mam again when I was a kid.

Driving further and further into the fields of County Durham, away from Sunderland, it felt like we were on an adventure. Second by second more stars were becoming visible until there seemed to be millions filling the inky black sky. After about an hour we turned off down a lane which led to a dead end, guarded by a barrier. Dickie turned off his headlights and the car slowed to a stop. I was nervous with excitement: here I was in the middle of goodness knows where, and I hadn't even known

Dickie for a year yet. Maybe he was an axe murderer? His door opened but the interior light didn't come on – he had sorted that out before we left. If it had, it would have thrown white light everywhere and hindered our eyes' adjustment to the dark conditions. I stepped out of the car into pitch blackness. I couldn't see anything ahead of me, but I could hear murmuring in the distance. Funny that – we hadn't spotted any other people or cars on the approach. My heart began to race and I felt myself tensing as I took blind steps forward, my arms outstretched like a zombie. I was just starting to get used to being on our own, and less on edge, when I heard the loud crackle of leaves to our left. Two large shadows appeared from out of nowhere. 'Areet?!' Even in the dark they must have seen me jump!

The ruffle of thick nylon jackets and the smell of coffee helped me to focus on the two men, who, after a minute or so, I could see better. One was crouched by a large telescope, flask in hand; the other was standing upright. He turned on his red torch as Dickie introduced us.

'This is Gary. He's the neighbour I've been telling you about – a mad keen astronomer. Gary, meet the SAS's finest.'

Dickie had mentioned he was a member of Sunderland Astronomical Society, and I had been dying to come along to one of their meets. They both said a warm 'Hello', and Don Smith, the older guy of the pair, who was maybe in his sixties, asked if I would like to take a look through his telescope. I shuffled over nervously, delighted to be invited into the fold, but anxious not to make a fool of myself. I studied the eyepiece to work out the magnification value. 'Ah, fifty times magnification, nice.' I heard Dickie let out a snigger and we both smiled; I was learning.

I noticed they had a few hairdryers with them, which seemed odd. I asked Don and he said it was for the dew. The most common hassle amateur astronomers have with their kit is water condensing from the atmosphere onto their lens or eyepiece. Don had fitted rubber caps to his telescope that would help prevent the moisture dampening his view of the stars, but if all else failed, nothing could beat your bedroom blower.

We were at Millshield picnic site, situated next to Derwent Reservoir in the foothills of County Durham. It was a wide-open gravel expanse surrounded by trees, which protected us from the wind and the glow of lights in the distance. It was completely secluded apart from the birds flying overhead and a few ducks landing on the water – the only noise we could hear was the occasional quacking against the glug of the reservoir. Behind us, the water glistened as it reflected the starlight and a breeze blew at the trees, which swayed one way then the other. With each gust, more stars would become visible, whilst others would disappear. Silhouetted against all of this were figures of people dotted about with telescopes, each looking up silently into the

vastness of the universe. It felt calm and almost medita-
tive, a hush like being in a cathedral, but instead of
looking up and seeing stained glass and murals, it was
even better: stars everywhere, as if they were observing
us. I knew I was somewhere special.

The next few hours passed more quickly than I could
imagine and it was soon 4 a.m. Don and his friend
treated me like one of their own, answering every question I had and showing me everything from the piercing
orange/yellow hue of Jupiter to the clearest Milky Way
arcing overhead I had ever witnessed at that point; each
star was so much brighter than I had seen from my
back garden. Don couldn't wait to tell me what he knew
and share his enthusiasm – it was infectious. I learned
that Don was a secondary school science teacher for his
day job, but like most of the amateur astronomers in
the group, his main passion was the sky. He was working
on new ways to educate kids and adults so that they
too could enjoy the stars. I liked him immediately; he
was patient and unassuming and a real gentleman. He
even made his own telescope, a feat that earned him
cult status in my mind.

Huddled under the starry sky with my new friends,
I didn't see the aurora that night, at least not with my
naked eye. But the SAS, as they called themselves, taught
me a few handy tricks. The first lesson was clouds. When
there are intermittent, bog-standard clouds in the sky,
they block out starlight completely, which seems obvious

enough; a cloudy night is an astronomer's nemesis. But the aurora is different: unlike clouds, starlight shines straight through the Northern Lights. Don and Dickie were telling me this because from mid latitudes like Derwent Reservoir, the aurora can often appear similar in appearance to ambiguous cloud-like formations.* The acid test is, can you see stars shining through the clouds? If you can, they may be a weak aurora starting.

Just before we left Derwent that evening, I found my shrouded stars, twinkling through what appeared to be a dusty haze. I called out to Don and Dickie and a few moments later they rushed over with an SLR (single lens reflex) camera and we took a 20- to 30-second exposure. It wasn't digital back then, so we had to wait to get the film processed, but with any luck when the photo was developed we would see the bright colours of the aurora. The guys explained to me how our eyes work very differently from a camera; we take a series of continuous exposures that our brain cannot stack to make one long picture, but a camera can. It can remove the ambiguity and produce breathtaking images far superior to those from our own eyes, particularly of something as fleeting as the aurora.

'That's all well and good, but if I take a picture of a cow, it doesn't make me a farmer, does it?' Don Simpson,

* If you live in Alaska or northern Europe, the Northern Lights do look more pronounced.

the chair of the SAS, had heard the commotion and come over to say hello and offer his two cents.

I laughed out loud. A tall proud man, wearing his USS *Nimitz* baseball cap that I would later discover seemed to be permanently glued to his head, he shook my hand warmly and launched into telling me the finer points of what constituted a real astronomer. It turned out that our group that evening, as well as the amateur astronomy community in general, was divided into the imagers and the observers: the imagers were mainly interested in astrophotography and taking stunning photographs; the observers were more excited by the experience of running a telescope across the sky and studying celestial objects and space itself. It's a playful rivalry, a bit like a rival gang mentality: the observers would say that imagers aren't real astronomers, and vice versa. I took Don's point, but I knew he was being tongue-in-cheek. Some of the imagers were clearly computer scientists and boy did they have some kit. The PCs and instruments they had at their disposal could achieve high-resolution images of the most distant galaxies.*

After just a few hours with the two Dons and the group, I knew I wanted to be part of this community. They were eccentric, yes, and a little nerdy, but they

---

* With the rise of digital cameras since the mid-1990s, astrophotography has grown enormously in popularity and the results that can be achieved with a DSLR now are extremely professional.

were dedicated to having fun and dreaming about what was out there, undiscovered, in space. They were hugely proud, too, of our heritage in the North-East. Don Simpson told me with reverence of the Venerable Bede, an eighth-century monk and scholar from Sunderland, who as well as being a great theologian and historian was one of the earliest astronomers of his time. From his small monasteries in Wearmouth and Jarrow, he wrote about how the world was round and how the Earth's tides were caused by the Moon; he was the first scholar to record these theories in writing, hundreds of years before gravity was discovered. Don told me, too, of Thomas Backhouse, a Victorian astronomer and meteorologist, again from Sunderland, who had discovered the first new star of the twentieth century in 1901. It was a momentous occasion for the local community and he studied the findings intensely with other astronomers from the area, making discoveries that would ripple around the world of science.

I wanted to be part of this great tradition; it really hooked me in. Not long after the trip I joined Sunderland Astronomical Society and began attending the meetings every Sunday evening at the Quaker's meeting house on the seafront. Arriving upstairs in the draughty second-floor room was the highlight of my week; I would chat, make new friends and plan our next dark-sky trip to go observing. For a few hours each week I forgot my working life on the building site entirely. The group was made up of all ages and backgrounds, and was well organised as well as being passionate and welcoming. I crammed in as many stargazing nights as I could manage over the first few years, and was soon getting involved in organising fundraisers and camps to promote astronomy in the area. Three of the closest friends I made during this time in the SAS were Don Smith, Jack Newton and Jürgen Schmoll. We bonded over a short trip to a new forest that had yet to be explored by astronomers. It would be the next step in my journey as an astronomer.

*

'Damn you, Newton.'

I'd heard a painful thud and knew instinctively someone very close by had tripped and fallen. I didn't know whether to laugh or run to offer assistance, my tired eyes adding to the general sense of delirium. In the dark I could just make out a slight figure, arms flapping as he patted the dirt off himself. I could hear muffled profanities in a foreign accent. There was only one person it could be: gradually the ponytail and the specs came into view. Imagine a German Robin Williams. Well, you have Jürgen Schmoll, the astronomer-cum-comedian of our star camp.

Moments earlier Jürgen had been trying to find his way to his tent after a night's observing, but had encountered a drop in the lay of the land and down he went, cursing Newton for his blasted gravity. I went to help and we laughed as we navigated him gingerly back to his tent.

'Would you like a fizzy yeast, Gary?' Jürgen asked, his breath condensing in the air.

'A what?'

He pulled an ice-cold can of beer from the six-pack cooling outside his tent, and chucked it through the air. I loved this guy. We had a few more fizzy yeasts before he turned in for the evening.

I had to stay awake for a little while longer as I was on duty as the organiser. The site was silent apart from an owl twittering away, and the grass was white with frost, crunching under foot with every step. I looked up again, for what must have been the thousandth time that evening, and was still awestruck. The sky was far darker and clearer than at Derwent, or anywhere I had been. I felt as if I could reach up and touch it, as if it were three-dimensional. The shapes of the constellations were not easy to spot; such was the volume of stars visible, they faded into a swarm of starlight. I hardly knew where to look, there was just so much to take in.

The sky finally began to change as the Sun was rising in the east, which was my cue for bed. I turned to walk to my tent and passed again where Jürgen was sleeping. As I approached him, I could see something sticking out of the bottom of the tent where the zip is. I stopped and stared. I couldn't believe it. It must have been 5 degrees below zero, and there were his feet – both of them – no shoes, just socks, resting in the wet frost.

'Jürgen?' I whispered. 'Are you OK?' I heard a muffled response.

'Ah, yes, Gary, I'm fine.'

'Mate, your feet are sticking out of the bottom of your tent.'

'Ah, my scopes are taking up a lot of room.'

I could hear him shivering now; his tent was tiny, maybe 3 by 1.5 metres and it was full of kit, so much so that he barely had room to sleep.

'Before you catch your death, do you want to share mine? There's room for six.' He accepted gratefully and we stayed up and talked for a few more hours. It was dawning on me that it wasn't all about the astronomy and the physics, it was about friendship.

*

Much to my relief, the first Kielder star camp had been a triumphant success. There must have been eighty pitches occupied for the two-day event at the end of September, and the campsite was buzzing. It felt like a space station on a distant planet: tents illuminated from the inside by red lights, quiet voices chattering away in dark corners interspersed with the whirring of telescope motors.

My inspiration for the camp had been Don Smith. A few years before, just after I joined the SAS, he had organised a stargazing trip for a group of us to a log cabin in a place where the skies were apparently even darker and clearer than Derwent. Kielder Forest was a large natural expanse that had been cultivated to allow for crops of trees to power the steel plants on Teesside during the Industrial Revolution. It is home to Europe's largest man-made forest, predominantly consisting of Sitka spruces and a reservoir. A ninety-minute drive from Sunderland, it was a trek of sorts into the wilderness, into a part of England that felt like Scandinavia. But its isolation meant that the light pollution was almost non-existent, and it was so far north that it could offer a vantage point for the aurora.

Up and down the country we knew there was a growing movement in amateur astronomy of darkness-seekers. Word had got around of an astronomy meet-up at Dower House caravan park in Thetford Forest, Norwich. It was called a 'star camp', and here like-minded astronomers could pitch a tent and spend a few days outdoors observing the clearest skies. I went down with Dickie and Jack and a few others. It was a glorious weekend of events and conversation, but one thing kept nagging at me the whole time we were there. On the journey home after the event, I brought it up with Dickie and Jack.

'Kielder skies are darker than those. I mean, way darker . . . We could see the Milky Way, which is great and all. But the Milky Way is so much clearer and more impressive at Kielder.' The lads could tell where I was going with this. 'What if we had a similar event at Kielder?'

'Nah, it wouldn't work,' said Dickie.

'I doubt it, Gaz,' said Jack. 'It's just too far north, no one would go that far.'

'I dunno. If we're anything to go by, these folks are pretty dedicated. Isn't it worth a shot?' My mind was racing and the first plans for the Kielder Forest star camp had been hatched. Little did I know then that in five years' time it would be voted one of the top ten star camps in the world, up there with the Texas Star Party, where the skies are advertised as being as dark as coal.

When I got back home, I just had a feeling it could work. If there was one thing I knew, it was how to work hard. It had already been drilled into my DNA through years of hammering away on building sites. I had never felt as enthused in my life as I did about this venture. I couldn't stop telling people and it didn't take long before the word got out that some bricklayer, some stargazer, wanted to do something with Kielder.

It was around this time that I met Pippa Kirkham, a ranger who was working for the Forestry Commission in Kielder Water and Forest Park. Pippa had heard about my efforts and we arranged to meet up. I liked her instantly. She was warm and friendly and told me about the themed nights she was running at Kielder Castle, a Gothic eighteenth-century building in the heart of the forest. The castle had been fully restored and was open to the public for these talks, but they wanted to try a different tack and move away from the ghost events that the castle had become known for. Originally built as a hunting lodge for the Duke of Northumberland, it had been constructed on a burial ground dating back to around 3000 BC. Instead of scaring guests with stories of its haunted servant girl, who was known to stalk the staircase late at night, Pippa was leading a campaign to promote 'astro-tourism'. Would I like to give a talk on astronomy? I couldn't be worse than a ghost-hunter.

It took me back completely at first. Me, give a talk? I didn't know whether I could pull it off. I still felt quite new to this world, even though I had been reading physics books most of my life and was now a capable observer. I also hadn't received much in the way of formal presentation training, although I was competent barking instructions to my colleagues on the building site.

Thankfully there was a press officer for the Forestry Commission, Richard Darn, who was also an amateur astronomer and would be on hand to help. The evening arrived, fortunately with a clear forecast, and Richard and I waited nervously outside the castle, scopes at the ready. As the punters slowly appeared, I was nervous. What on earth was I going to say? What if they didn't like me? What if they thought I was a fraud? Pippa herded the group towards me and everything went quiet as she started my introduction. She looked around the crowd and paused. 'This is Gary, he is an astronomer.'

With these words, every doubt dissipated. Astronomer. I was accepted; I was electrified into action. That evening I pointed out the North Star and the satellites orbiting the Earth to the visitors. I spoke of the aurora, and thought back to the first time I saw it with my dad. I shared it all and I felt elated; the audience responded.

Six months later, and after several more talks at Kielder Castle, I was standing on the campsite on the final day of our first Kielder Forest star camp, just a five-minute walk away from the castle and close to the Anglers Arms pub for refreshments. I had organised the event with the help of Pippa, Richard and the Forestry Commission. In attendance were a hundred visitors, as well as expert speakers giving lectures and astronomy vendors selling the latest telescopes and kit. I had awoken bleary-eyed from lack of sleep, but fuelled with excitement. Jürgen, Jack and I walked over to the shower block for a warm shower, where we overheard visitors standing around, discussing the night before. 'That galaxy was bright . . . I've never seen it as bright as that before, not in the UK.' We smiled to each other, delighted to hear the chatter. We were on to something and my mind began racing again. What if we had a more permanent camp?

Just as my thoughts were running away from me, Richard wandered over. He had a smile on his face. 'Gary, a BBC TV crew are coming from *Look North* to do an outside broadcast for the weather. They want to interview you.'

My stomach twisted and time seemed temporarily to pause. My thoughts turned to Einstein and I remembered a quote of his: 'an hour with a pretty girl seems like a second, and a second sitting on a hot plate feels like an hour, that's relativity'. I re-ran what Richard had said, and somehow instantly knew what the repercussions of a live TV appearance would be for me. Not fame or fortune; this was local news, after all. But my passion would finally be out in the open. The football lads, everyone who I had kept my astronomy quiet from, would probably find out. Was I ready for the questions? To become the 'Curly Watts' of Sunderland? I did the interview and gave it all I had; it was liberating. It seemed as if the final weight had been lifted from my shoulders. Although I still felt a bit like an intruder – I didn't have the formal education or a degree – that night I came out, and I wasn't going back, ever. My future was firmly among the stars.

May–June Sky Guide

Leo – Virgo – Boötes – Hercules – Venus

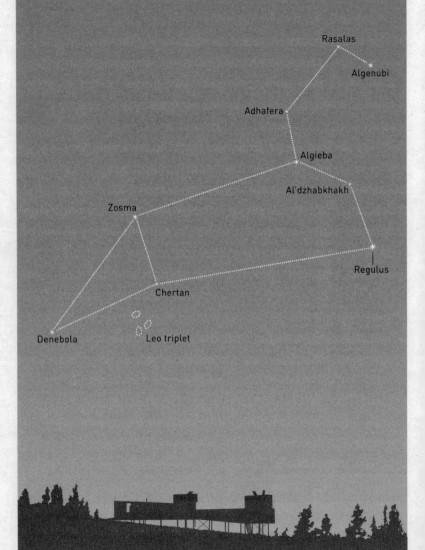

LEO

In summer (June) looking west from Kielder Observatory

Rasalas

Algenubi

Adhafera

Algieba

Al'dzhabkhakh

Zosma

Regulus

Chertan

Denebola

Leo triplet

Stars   Mag 0 ✷   Mag 1 ✸   Mag 2 ✦   Mag 3 ✲   Mag 4 ·   Mag 5 ·   Clusters ⁚   Nebulas ☐

As one of the twelve zodiac constellations, Leo sits on the ecliptic. Traditionally seen as a spring constellation, by early May at approximately 9 p.m. you'll find him sitting due south, underneath the feet of Ursa Major. By late June he is setting towards the west.

People often remark that the constellations don't look like the characters they are supposed to represent. Leo, however, is one of the most realistic with its stars appearing like an Egyptian sphinx with its front paws outstretched. The most easily recognisable stars in the constellation are those in the lion's head and chest as they form a backwards question-mark shape known as the Sickle. The full stop at the base of the question mark is Leo's brightest star – Regulus. You should see it as a double star through binoculars. It has been suggested that a conjunction between Regulus, the king of the stars, and Jupiter, the king of the planets, is what the Three Wise Men referred to as the Star of Bethlehem.

From Regulus, run along the lion's underbelly and you'll find Chertan (θ Leo) at his rump and Denebola (β Leo) slightly further along in his tail. That tail joins the back of his body at the star Zosma (δ Leo) and his back runs along the line between Zosma and Algeiba (γ Leo).

Scattered around this framework of stars is a treasure trove of distant galaxies. Start from Denebola in the lion's tail and once you reach Chertan take a sharp 90-degree turn clockwise. You're looking for the magnitude-5.3 star η Leo. Now take another 90-degree clockwise turn and continue in a straight line for a little less than a degree. You have now entered the realm of the famous Leo Triplet – three galaxies at very close quarters. Heading in from η Leo, you'll encounter M65 first, quickly followed by M66. The two galaxies are gravitationally bound together with the third galaxy in the Leo Trio – NGC 3628. It can be found by taking the midpoint between the first two galaxies and heading slightly back up towards the underbelly of Leo.

Of the three, M66 is arguably the most spectacular as its spiral arms are far more vibrant. It is thought the gravitational interaction between the galaxies has enhanced star formation within it. M65, on the other hand, seems to have been less affected as its own arms are dull by comparison. Whilst the first two are face-on spirals, we are seeing NGC 3628 edge-on and so its structure is harder to make out. As all three galaxies sit around magnitude 10, you'll need a fairly large telescope in order to observe them. These galaxies, whilst they are telescopic objects, are an absolute must for observers as they progress; think of locating them as passing an observer's examination.

Our galaxy-hunting is far from over, however. Return to Chertan and head along the line towards Regulus. Just before halfway you will find the magnitude-5.5 star κ Leo (sometimes known as Al Minliar al Asad). Turn off at a right angle again, down beneath Leo, until you reach the star HIP 52683. Continuing past it for almost the same distance again will bring you to M96. Whilst it is about the same size and mass as our Milky Way, it is very asymmetrical. Again, this is thought to be as a result of gravitational interactions with neighbouring galaxies. One of those neighbours – M95 – sits only half a degree away, up and slightly to the right. However, at magnitude 11.4 it is harder to make out.

If you imagine M95 as the right-hand corner of an inverted triangle with M96 at its tip, then the galaxy M105 sits at the corresponding left-hand apex (less than a degree away). Rather than a spiral, it is an elliptical galaxy shaped more like a rugby ball than a flat disc. Elliptical galaxies are thought to rotate a lot more slowly than their spiral cousins, if at all. A difficult test for sure.

VIRGO

In summer (May) looking south from Kielder Observatory

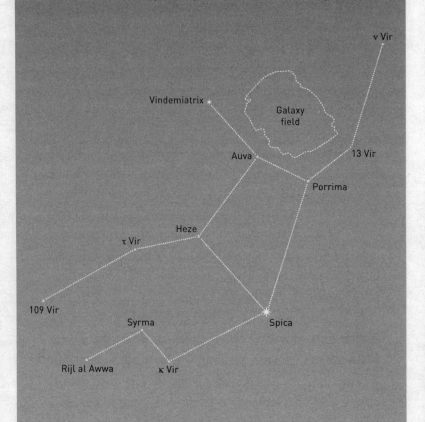

ν Vir

Vindemiatrix

Galaxy
field

Auva

13 Vir

Porrima

Heze

τ Vir

109 Vir

Syrma

Spica

Rijl al Awwa

κ Vir

Stars   Mag 0 ✸   Mag 1 ✴   Mag 2 ✦   Mag 3 ✶   Mag 4 ·   Mag 5 ·   Clusters ⁘   Nebulas ☐

The shape of the constellation Virgo is depicted as a young maiden holding a sheaf of wheat in her hand. The sheaf is represented by the star Spica, the brightest in Virgo, and finding Spica is the easiest way to locate yourself in this sprawling constellation – the second largest in the night sky and trumped for size only by the much fainter Hydra. You'll quickly enter Virgo if you start at Zosma in Leo and draw a line through and past Denebola. A far more common way to navigate is to use the Plough in Ursa Major and Arcturus in Boötes (see next constellation).

Move up in the sky from Spica and you'll reach Heze (ζ Vir) in her leg. Taking a 90-degree turn to the right will lead you to Auva (δ Vir) in the middle of her chest. This star is flanked on either side by the beautifully named arm stars of Vindemiatrix (ε Vir) and Porrima (γ Vir). Virgo's second brightest star – the fantastically named Zavijava – sits just to the right of Virgo's head, approximately halfway between Spica and Regulus.

Virgo is home to the mighty Virgo cluster of galaxies. Just as planets gather into solar systems and stars combine into galaxies, so galaxies themselves clump together under the attractive force of gravity. Our own Milky Way belongs to a set of galaxies known as the Local Group. In turn, clusters huddle together into vast cosmic structures known as superclusters. It is from these superclusters that the universe itself is constructed. Both the Local Group and the Virgo cluster sit within the enormous Virgo supercluster. But whilst our Local Group contains around sixty galaxies, the Virgo cluster contains at least 1,300. The brightest can be seen in the direction of this constellation.

The giant elliptical galaxy M87 is a particularly big beast, with a mass double that of our own Milky Way. It also has over 12,000 associated globular clusters compared to our lowly 150. All galaxies are thought to have a central black hole and our galaxy's black hole weighs in at 4 million times heavier than the Sun. M87's, however, trumps that at a colossal 7 billion solar masses. Astronomers have noted an enormous jet erupting from the centre of M87 which is thought to be associated with that black hole.

You can spot M87 for yourself in the northern part of Virgo, along the line between Vindemiatrix and Denebola in Leo. It sits just over a third of the way along, just above the star HIP 61135. It is in this region that you will also find a multitude of other galaxies. You can pick from M49, M58, M59, M60, M61, M84, M86, M89 and M90, which vary between spirals, ellipticals and lenticulars (the latter are part way between an elliptical and a spiral).

These are difficult to see through binoculars, but through a medium-aperture telescope they are breathtaking.

However, Virgo's most famous galaxy among amateur astronomers – M104 – is not part of the cluster at all. More commonly known as the Sombrero Galaxy because of its resemblance to the Mexican hat, it appears that way because it is an edge-on spiral galaxy. You can track it down by moving over to Porrima and finding the midpoint of the line down to Spica. From here, take a right-angled diversion towards the star Algorab (δ Cor) in Corvus, the Crow. Before you get there, you should encounter the star HIP 61656, which sits at the top of three stars all in a line. Skip from one star to the next away from HIP 61656 and continue onwards until you have the Sombrero in your sights. Its own central black hole has a mass equivalent to 1 billion suns, making it one of the largest black holes known in the local universe.

BOÖTES

In summer (June) looking south from Kielder Observatory

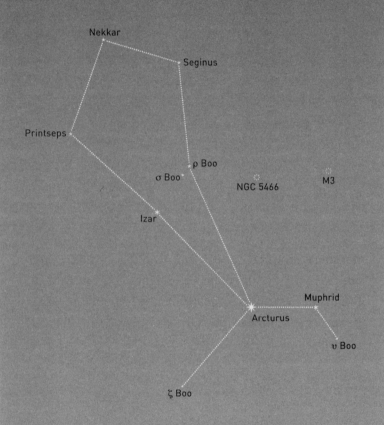

Nekkar

Seginus

Printseps

ρ Boo

σ Boo

NGC 5466

M3

Izar

Muphrid

Arcturus

υ Boo

ζ Boo

Stars   Mag 0 ✹   Mag 1 ✸   Mag 2 ✶   Mag 3 ✦   Mag 4 ·   Mag 5 ·   Clusters ✺   Nebulas ▢

The constellation of Boötes is drawn as a herdsman, and is often referred to as the 'bear-herder', as he is seen to follow the constellation of Ursa Major across the sky. Some say he is Arcas, whose mother Callisto was transformed into the constellation of the Great Bear (Ursa Major). Like a good son, perhaps he is simply watching out for his mother.

The unmistakable red star of Arcturus is the jewel of this constellation. Shining with a magnitude of -0.05, it is the second brightest star in the northern hemisphere sky (the fourth brightest over both hemispheres). There are two reasons for its brilliance: it is inherently very luminous, but it is also relatively close to us. The traditional way to locate it is to start with the Plough asterism in Ursa Major and follow the 'arc' of the handle to Arcturus. Continuing that curve onwards allows you to speed on to Spica in Virgo.

Arcturus sits at the base of a distinctive kite shape which forms the body of the Herdsman. Travelling up and to the left will take you to Izar (ε Boo) and then Princeps (δ Boo), before you reach his head at Nekkar (β Boo). Returning to Arcturus down the opposite side of his body sees you pass through Seginus (γ Boo) and Hemelein Prima (ρ Boo). The star Murphid (η Boo) sits just to the lower right of Arcturus and forms Boötes' leg.

Boötes is an area rich in double stars, with Izar being a particularly fine example. However, the two stars are relatively close together so you'll need a telescope with an aperture of about 3 inches in order to split them apart. Should you do so, your reward will be a duo exhibiting beautifully contrasting colours. Other doubles to look out for in this region include π Boo and ξ Boo.

Double stars may be plentiful here, but Boötes is fairly light on other deep-sky objects, particularly those visible without a telescope of sizeable aperture. This is because the constellation resides far from the dense band of the Milky Way. There is one globular cluster to hunt for, but those wishing to see these compact stellar conglomerations are perhaps better off looking at Hercules, our next constellation. Boötes' cluster – NGC 5466 – shines with a lowly magnitude of 9.1, so you're starting to stretch a pair of binoculars to their limit. Its lack of brightness results from it being fairly far away (over 50,000 light years) and its stars not being as tightly packed as other globular clusters. Those with fairly large telescopes can find it by starting with Hemelein Prima (ρ Boo) and heading off on a straight line towards the magnitude-5.0 star HIP 68103.

There are a couple of galaxies on offer here, but none to match the spectacles in Leo and Virgo. At magnitude

11, NGC 5248 is the brighter of the two and is found by starting with Izar and drawing a line straight through Arcturus towards Virgo. However, the dimmer NGC 5676, located above Boötes towards the handle of the Plough, is arguably more interesting. For starters, it appears not to have a central bar structure in the way that our own Milky Way does. Also, unlike our galaxy, its spiral arms are disordered and fragmented. Astronomers refer to this kind of chaotic galaxy as 'flocculent'.

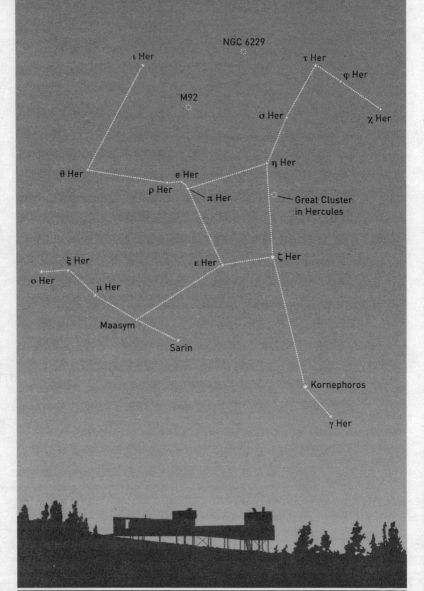

HERCULES

In summer (June) looking south-east from Kielder Observatory

NGC 6229

ι Her

τ Her

φ Her

M92

σ Her

χ Her

θ Her

e Her

η Her

ρ Her

π Her

Great Cluster
in Hercules

ξ Her

ε Her

ζ Her

o Her

μ Her

Maasym

Sarin

Kornephoros

γ Her

Stars   Mag 0 ✳   Mag 1 ✷   Mag 2 ★   Mag 3 ★   Mag 4 ·   Mag 5 ·   Clusters ☼   • Nebulas ☐

Once again we encounter an illegitimate son of Zeus. At the beginning of May, Hercules, which resembles the mythical Greek strongman, is rising in the east as the Sun sets. By the end of June he is found almost due south at midnight, just to the left of Boötes.

Strangely for such a revered character, there are no stars of first or second magnitude in Hercules. At first glance he seems to have been assigned a fairly unremarkable part of the sky. His most distinctive feature is the asterism known as the Keystone, which constitutes his body. Four stars in a slightly squashed square, they resemble a test-tube bung. Starting in the bottom right-hand corner, you'll find the otherwise nameless star ζ Her. Cross the base of the Keystone to Cujam (ε Her), before climbing the left-hand side to π Her and skimming across the top to η Her.

From each of the four corners of the Keystone spring his arms and legs (it is worth noting that seen from the northern hemisphere he appears upside down with legs upwards and arms downwards). Hercules' two brightest stars are found near the limb originating with ζ Her. At the end of that arm you'll find Kornephoros (β Her), with Rasalgethi (α Her) sitting just below his head near the bright star Rasalhague (α Oph) in neighbouring Ophiuchus. As is sometimes the case, despite its Bayer designation, Kornephoros is brighter than Rasalgethi. Through a small telescope you will quickly realise that the latter is a double star with separate red and blue components.

Hercules is home to the northern hemisphere's brightest globular cluster – M13. It is located approximately a third of the way along the line between the stars ζ Her and η Her on the right-hand side of the Keystone. With a magnitude of 5.8, it sits right on the cusp of human eyesight, so on a clear night you might just be able to make it out with the unaided eye. You should have no problem finding it with a pair of binoculars, however, and through a small telescope you should be able to resolve some of its 300,000 individual stars. In 1974, astronomers transmitted the so-called 'Arecibo message' towards M13. A radio signal, it contains information about the human race and our location within the Milky Way. However, the cluster sits just over 25,000 light years from Earth and so the signal isn't even close to arriving yet.

Another, less well-known globular cluster – M92 – sits not far above the Keystone between the stars HIP 83947 and HIP 84850. Appearing slightly dimmer than its neighbour at magnitude 6.3, it resides slightly further away from the Earth too. Its constituents are some of the oldest known stars in existence, and are almost as ancient as the universe itself. The cluster lies on the circle traced out by the Earth's North Pole as it wobbles around due to precession. So, around 10,000 years ago, M92 was where Polaris is now and we had a pole cluster rather than a pole star. This would easily be the best globular in our skies if it were not for its neighbour M13.

Venus

Venus is unmissable. Blazing with a maximum magnitude of nearly -5.0, it is the third brightest natural object in the sky after the Sun and the Moon. This is partly because it is the closest planet to us, but also as a result of Venus' thick, highly reflective atmosphere. Venus' clouds – almost exclusively made from carbon dioxide – reflect around 75 per cent of the light falling on them. By contrast, Earth only reflects 30 per cent of sunlight back into space.

As the second planet, it never strays very far from the Sun in our sky. This means that we are only able to see it for a few hours around either sunrise or sunset. It is often labelled the Morning Star or Evening Star for this reason. At its greatest elongation (maximum apparent separation from the Sun), the gap between the two objects can reach a maximum of 47 degrees. That's a considerable improvement on Mercury, which never appears more than 28 degrees from the Sun. Venus can either be at 'greatest elongation east', in which case it appears in the evening sky, or 'greatest elongation west', when you can see it in the morning.

If you look at Venus over several weeks, you will notice that it exhibits phases just like the Moon. This is because it orbits the Sun at closer quarters than us and so the amount of sunlight it reflects towards us changes. When Italian astronomer Galileo Galilei first noticed this at the beginning of the seventeenth century, it was the first nail in the coffin for the idea that the Earth was at the centre of the solar system. Only by placing the Sun at the centre – with both planets orbiting it – can you account for why Venus appears to change appearance in this way.

The best time to watch the planet's changing phases is as Venus moves between greatest elongation east and inferior conjunction (when it sits between the Earth and the Sun). It will start approximately half illuminated, but will slowly shrink into a thin crescent over the course of about ten weeks, before disappearing from view completely as it passes close to the Sun. It will then reappear in the morning sky and grow in phase as it heads towards greatest elongation west.

Here's a table that shows at a glance when the planet is due to reach these points in its orbit in the years ahead.

Greatest Elongation East	Inferior Conjunction	Greatest Elongation West
12 January 2017	25 March 2017	3 June 2017
17 August 2018	26 October 2018	6 January 2019
24 March 2020	3 June 2020	13 August 2020

During most inferior conjunctions, Venus is seen to pass above or below the disc of the Sun due to Earth and Venus' differing inclinations. Occasionally, however, Venus passes directly in front of the Sun from our perspective. This incredibly rare event is known as the Transit of Venus, and it last occurred in 2012. It will not be seen again until 2117.

4

Mothership

'Ahhh-choo!' My sneeze ricochets around the half-finished classroom, disturbing the silence and the sawdust. It is mid-evening on a Friday in late April 2008 and the Moon is rising. I'm weary from a day's work on the building site, but happy to be out here now, as always. I can hear the trees outside creaking in the wind and the smell of fresh timber and machinery is pleasing – the smell of construction.

I'm alone, but sawdust cushions my every step as I creep around the soon-to-be-completed Kielder Observatory. Like everything about the project, the all-timber structure has ballooned in size and possibility since we chose the London architect Charles Barclay's stunning design over a year ago. The computer, chairs and telescopes haven't arrived yet and the log-burning stove hasn't been installed, but this dark room, which will soon comfortably house forty people for stargazing lessons, is nearly ready. Like the rest of the observatory, there are no windows; we've designed each room to be an oculus. The only way you can look at the sky is to peer through the telescopes and out into the universe.

Only a few wooden planks still need to be laid down. We've chosen Siberian larch, a tough and long-lasting wood as strong as oak, but sourced from sustainable forests. It should still look good in twenty years' time; its panels robust even after I'm gone. Despite my years as a builder, I'm no carpenter, so fortunately I'm not contributing to the labour this time around. Two days ago I watched in awe as the five master carpenters toiled away, sculpting the timber structure one beam at a time. Shoulders hunched, eyes focused, they were working a ten-hour shift despite the freezing rain and winds. I was amazed by their level of commitment – real grafters – but also by their level of skill, which verged on the artistic. They flawlessly planed and assembled the walls, roof and floor so that it was perfectly smooth, with no gaps in the wood or splinters. It was as if the structure was creating itself organically, and maybe it was: one of the reasons we chose this design was that it looked like an otherworldly spaceship resting on the hill, sailing out into the night.

Tonight I approached the small gate at the bottom of the track with a single thought. How many times would I be doing this in the coming years? The modernist, low-slung facility rose up in front of me. Built on wooden stilts, which are more pronounced at the southern end to counteract the slope of the hill that overlooks the Kielder Water and Forest Park, it is ostensibly a pier on land: the classroom and the two square turrets that will house the telescopes separated by a viewing deck. The turrets, an unusual design choice, have clean, straight lines which suggest a scientific purpose, but stop short of revealing the observatory's true intent. During the day, it will double as a viewing outpost that can be visited by the general public; we didn't want the white or silver hemispherical domes, which are used for most astronomical observatories, disturbing the landscape.

I walk out onto the deck to survey the land in the twilight. Everything is mostly quiet, apart from the whistling of the wind turbine, which will help to power the telescope's computers along with the solar panels and the generator. When the wind is still, you can discern on each of the three blades of the turbine two painted words. Together they form a circular poem which has been composed especially for the observatory by the poet Alec Finlay:

space arcs
light eclipses
time bends

I suspect the words may eventually peel off in bad weather, but I know their meaning will last. As I rotate the poem around in my head, I look south towards the reservoir and the valley. At an altitude of just over 330 metres above sea level, the observatory is situated on one of the highest peaks in the Kielder Park: the view of the horizon is unhindered in all directions. This elevation has practical benefits. Foggy conditions, which can occur as a direct result of the valley beneath us, on top of moisture from the reservoir, can cause condensation on our telescopes. However, up here these effects are reduced. Another benefit is that the observatory encounters less of our atmosphere's distorting effects, which can make stars twinkle and become harder to see. But more simply, the vista is a thing of beauty.

Perched on Black Fell, the observatory is surrounded by hundreds of square miles of untamed wilderness. Off-grid and almost entirely absent of people, elegant Sitka spruces and wild scrub populate every inch of the rolling hills. The closest neighbours in the area are probably the scurrying red squirrels, of which we home nearly half of England's population. Joining the silent party on land are roaming herds of wild deer and goats; soaring above us is a wake of buzzards and the occasional goshawk.

Still standing on the viewing deck, I turn my head southerly towards the tranquil blue of Kielder Water in the distance. If you're not from the north of England, you probably wouldn't know that beneath the water lies a village, or that in the mid-1970s forty families were made to pack up their belongings and evacuate their homes so that the valley could be flooded. In 1982, the Queen cut the ribbon to open Kielder Reservoir, declaring it the largest man-made body of water in the UK. Built to serve the burgeoning factories and thronging populations of the north, sadly the vast lake became a white elephant. The economy stuttered and the mines and heavy industry failed. It remained full but largely unused, though it was rumoured to contain enough water to flush every toilet in the world once, a fact I have yet to see substantiated, although maybe one day I will do the maths. Despite the controversy, I have grown to love the way the lake provides a focal point for the valley's wild, rugged topography.

As the temperature outside begins to drop even further, I head back inside the observatory, this time to inspect the square turrets that will house the telescopes. The scopes will be in the centre of each room on concrete pillars and supported by huge concrete pads in order to eliminate any vibrations and ensure that observations are as smooth and sensitive as possible. To take advantage of Kielder's dark skies and lack of light pollution, we decided that there would be two permanently mounted telescopes. One is to be a 14-inch Meade telescope that was designed to be an all-rounder capable of performing standard imaging – the planets, the Moon and closer stars – as well as observing deep space. It is fitted with a GPS system for ease of use. The other telescope has a 20-inch aperture whose greater grasp of light will enable us to see distant objects with better resolution; its superior aperture will grant us unprecedented views. We hope this second scope in particular, combined with Kielder's jet-black skies and the promise of the aurora, will become a crowd-puller for amateur astronomers and curious minds from up and down the country.

The observatory is taking shape now about a year since the first foundations for the site were dug. I inspect the four newly installed metal shutters built into the turret which will open from the inside, revealing a portion of the sky for each telescope to be trained on. An important feature of each turret is its ability to rotate 360 degrees, so that each telescope can be trained anywhere in the sky. Because we can't get mains electricity up here, we've had to rule out any chance of having motors turn the turrets, which weigh nearly 6 tonnes each. The only feasible alternative is to rotate them with manpower – by hand. The upper portion of each square turret was therefore built on a circular metal ring, and a large gearbox was fitted with a steel handle on a big wheel. Its design is precision in essence, and it's very effective and fun to use: turn the handle and it turns the gearbox, which imparts its energy onto the ring. Finally, this rotates the turret so that the observer can point the telescope wherever they like. With all of the complex technology that will be functioning at once inside the observatory, it's comforting to know that a big wheel with a handle on it actually makes it all possible.

I sit on the floor and begin to daydream about the observatory once it is open. I imagine sitting under the scope and playing Pink Floyd through the speakers . . . pouring myself a single malt . . . looking at undiscovered star clusters late into the morning . . . when my reverie is interrupted by the noise of a car pulling up in the gravel drive.

It is Steve Mersh, a contractor from the Home Counties who has led construction on the site. Steve is of slight build, but graced with a steely determination and resolve I admire; quite the type I would expect to meet on a building site. He and his team have needed all of that resolve during the cold, harsh Kielder winter; I can hardly believe they have worked through it all and we are now so close to the finish line.

I call out to Steve and see that he is with one of his builders, Gavin, a burly Geordie from Bellingham who I've also got to know personally on the site. During one of my visits a few months ago, I entered the building as he was putting the finishing touches to the passageway. He was sat hunched by a bench, using a loud circular saw, dust flying everywhere.

'Hi Gavin, you OK?' I mouthed above the noise.

'So long as I keep moving, Gary. If I stop I will freeze up.'

I remind Gavin of that night now and he grimaces. Thankfully the freezing nights are nearly over.

A few minutes later I'm outside with Steve when I look over his shoulder. The crescent Moon has risen higher since I have been here, its light scattering through our dusty atmosphere, casting a yellow hue. The view is due east, and I can see the craters on the surface. Around it, stars like tiny light bulbs are switching on and stretching out in all directions, tracing shapes in the sky. As my eyes adjust, I notice one particularly bright star that seems to be zooming across the horizon. I have never seen it so bright before, but of course it isn't a star. It is one of man's greatest inventions: the most sophisticated laboratory ever built, orbiting our planet at nearly 8km per second, with some of our finest scientists and astronauts aboard – the International Space Station.

I find myself transfixed and staring up, imagining what they must be thinking, looking back down on us. This is why I am really here: not to inspect the building, but to check in with the cosmos. This is why we have worked so hard for so many years to raise the money and realise this dream – all to look up and marvel. A jolt surges through my body. This is what it's going to be like from now on, I think. My arms and back tingle. And this is just a glimpse of what a starry night at Kielder Observatory will truly be like. It is awesome, in the original sense of the word. And it is only the beginning. Are you ready? I think to myself.

*

The road to building the observatory at Kielder really began in an old pub next to my original dark-sky site, Millshield, near Derwent Reservoir. Ever since my first Kielder star camp I had wanted to build something under these dark skies. After two years, the star camps and Kielder Castle talks had blossomed into a regular feature, and I was buoyed up by the public interest. At the same time I was building the Cygnus Observatory for the Sunderland Astronomical Society. It didn't take long before word of our plans slipped out and ears started to twitch. Still, because of the forest's isolation, erecting any building there seemed like an insurmountable task.

I had been talking to a few people about raising maybe £1,500 to build a simple brick-based structure and cover it with another mossy plastic dome that had been donated. But looking across from me one evening in the pub, with a beer in his hand, was Peter Sharpe, the curator of art and architecture in Kielder. He had a much more ambitious plan. What if the observatory at Kielder could also become a leading work of architecture?

Since the turn of the millennium, art and sculpture have become central features of the Kielder Water and Forest Park. The region's crystal-clear lake and rich woodland has become an inspirational canvas for some of the world's finest artists. Nearly forty new public sculptures and artworks have been commissioned, including the futuristic Belvedere shelter, made from glimmering stainless-steel sheets that reflect the trees and light around them; the Kielder Column, a spiralling chimney of pink sandstone that reaches to the skies like a Gaudi-esque tower; and the wooden Silvas Capitalis, also known as the 'giant forest head', which you can walk inside via its surprised open mouth, and then peer out into the woods from its tribal eyes.

Peter's dream site for an observatory, although he didn't let on at first, was somewhere near the most famous 'sculpture' at Kielder, which had been built in 2000 by the Californian artist James Turrell. Skyspace is situated at Cat Cairn, a stony outcrop overlooking the valley. A circular dry-stone chamber, the building is semi-subterranean and you enter via a short tunnel dug into the side of the hill. Once you are inside, the walls are whitewashed and you can either take a seat or lie down on the bench that hugs the circumference of the room. Like a mini, primordial Pantheon, from there you look up and out through the oculus, the elliptical hole in the ceiling being the only source of light in the chamber. In the daytime this sliver of sky flitters from spotless blue to blinding sunlight to trundling Northumberland clouds. At night, it reveals a portal to the stars. By focusing on such a small section of the heavens, the sense of motion, as well as the colours and detail, is radically heightened. You couldn't feel further away from civilisation, which is why thousands of tourists visit every year to experience the strange, contemplative sensation.

Peter is a gentle, well-spoken man and he was warm and patient with me as he explained his ideas. He asked me about astronomy and the appetite for it in the region; I answered him honestly that it was growing and growing. He was desperate to produce something big that would really make a difference, and I could tell his passion for art matched mine for the stars. We spent most of the evening talking in hypotheticals, but as we were leaving Peter asked me what the estimated costs would be for the telescopes. I told him maybe a few thousand. He replied, 'Well, we should be planning on around about £150,000 as a budget for build and telescopes.'

My eyes nearly popped out of my head. How would we possibly do it? But on my journey home I could hardly contain my excitement. What would it look like? When could we build it? Would I be involved in the way I wanted to be? The following day the phone rang whilst I was at work on the building site. It was Peter and he left a message:

'Hi Gary, just a quick one to say thanks for last night. It was great to meet you finally. One thing we didn't mention was would you like to be the lead astronomer and founder of the project?' I didn't hesitate; we were off.

Within months of the initial meeting, our small, beery idea had snowballed into a steering group and plans to raise the funds. Armed with the frenzied enthusiasm of those trying to do something for the first time – and almost every step on this journey was into the unknown – we held fundraising meetings and put on more events at the Castle to raise awareness. We courted banks, businesses, councils and even European Development Funds, whilst the Forestry Commission helped us to scout out potential sites.

From the outset I wanted us to build a warm, safe observatory that would provide shelter and discovery for all ages and backgrounds – not a research site for professionals or academics. The aim was for a small team of astronomers, led by me, to put on four to six events a year at the observatory, teaching the general public about astronomy and the constellations.

With the help of Graham Gill, a softly spoken Scottish forest manager from the Forestry Commission, whose own love for astronomy has helped so much, we decided on Black Fell, a clearing in the wilderness half a mile up a forest track from James Turrell's Skyspace. We knew that the observatory would have to point south. The spinning of our planet produces the observed effect from Earth of stars rising in the east and setting in the west. As they track across the sky, they naturally have a point where they are positioned at their highest. If you draw an imaginary line from the North to the South Poles, we can call this line the meridian. Stars are always at their highest point in the southern sky sitting on the meridian – hence our observatory points south for the best view. But apart from the orientation, we had no inkling of what the observatory would really look like.

It was Peter Sharpe's idea to hold a competition. Instead of following our own noses with the design, we worked with the Royal Institute of British Architects (RIBA) to launch an anonymous competition, in which any architecture firm in the world could enter and submit a design for the observatory. We hoped we might attract a few imaginative entries due to it being quite an unusual brief: the chosen design in the heart of Kielder should have two telescope housings and a 'warm room', which would be large enough for adults to sit, work and stand.

Much to our surprise, though, we were inundated with responses. We received over 260 entries from across the globe. Soon we hardly had enough room in our team headquarters to house all of the submissions. I recall one design in particular that stood out. The concept drawings showed raised decks with bubble structures that seemed to be made out of glass or plastic. Each bubble contained a telescope, and was joined horizontally with glass tubes. It reminded me of Roald Dahl's

Charlie and the Chocolate Factory in which Augustus Gloop is sucked along the pipe. The proposed 'warm room' had open fires and was the highest in the building. It would be accessed from ground level via some sort of lift, or tube, which seemed to suck you up to the next level. On closer inspection, I realised to my disappointment that there would not be a suction device – I was hoping they had invented reverse-gravity tubes. So we passed on that one.

Another entry suggested we build a hut in a tree, complete with ropes and pulleys . . . We passed on that one too. One of the more practical designs was to be constructed in sleek black steel, with diagonal sheets that jutted to a point in the sky. I liked that design a lot; it also had a sauna in the warm room. A sauna? Yes, I had to do a double-take, but I wasn't seeing things: the drawing clearly indicated a person with a bare chest and towel around their waist. The concept sketch also showed a curious chap stood by the side of the building, with a flat cap on. Now, I'm not one for moaning about stereotypes and Kielder being 'up north', but I felt that adding a token miner to set the northern scene was taking the biscuit. Still, the design was shortlisted in the final six.

Hailing from Holland, Germany and across the UK, the shortlisted firms were interviewed in Newcastle over the course of a long day. All of the entries were excellent, but after hours of testing their concepts, a winner emerged. Alas, it wasn't the crew with the spiritual miner. There was only one design for me; only one architect who truly 'got it'.

*

25 April 2008
'Will this do, Gary?'
I give Charles a wry smile. Charles has probably aged as much as I have during this process. Over the last two years there have been thousands of tiny details to discuss, hurdles to jump and probably just as many phone calls to make. But I have always loved his clean lines and geometric design for the observatory. The budget has grown to nearly half a million pounds; bigger than we ever could have imagined. Now it is finally finished. We are both dumbstruck at its beauty.

Before the ribbon-cutting exercise I have to don a shirt and tie and show up at Kielder Castle with all the local politicians, of which there seem to be many. I hardly know any of them, but nevertheless the place is crammed like a jar of onions in vinegar. I chat to people and exchange small talk, but I feel very awkward; I just want to be at the observatory. All of that energy just waiting . . . my own personal star gate . . . why the hell would I want to chat to this lot? I mean, hardly any of them know the difference between a haemorrhoid and an asteroid. But that's just my defensive voice speaking, and my nerves. I'm eager to get the public astronomy programme moving – for ordinary people to experience what I have seen.

At 8 p.m. the wait is over. I look around and pause, take a few short breaths.

'Good evening, ladies and gentlemen . . . I'm not sure what is going to happen tonight . . . but welcome to the opening night of the Kielder Observatory.'

There is a smattering of applause and a few yelps among the smiling crowd. Everyone seems determined to have a good time. On this, the very first night, I've called on Richard Darn, my friend and fellow amateur astronomer, to be my lifeline, to help get me through it. I know he has my back and we've worked together countless times before. We grin nervously at each other as we welcome the forty guests. The event was advertised as a beginner's guide to stargazing and I've prepared a short talk entitled 'The Big Universe'. People have been arriving over the last few hours in their fleeces and warm coats: some with flasks, some with star charts and some with binoculars or their own portable telescopes. Many have been enjoying standing out on the deck for the last half-hour, perching on the wooden ledge, gazing out at the Kielder valley below. Most families have driven from towns and cities across the region and are making the most of this chance to be outside in the boundless countryside. Others have come from even further afield and are checking their star charts for the night's obser-vation. One couple were monitoring the aurora forecast before they arrived. Although the chances are low, they remain optimistic. I keep expecting the SAS contingent to arrive, the motley crew running up to give me big bear-hugs; but sadly they couldn't make it tonight. Soon. It is too late for my children, as well. Soon.

Once everyone is inside, I walk to the front of the classroom; I can hardly speak. My throat is dry and I feel my eyes on the verge of welling up. Behind me a large screen is displaying a picture of the Orion Nebula. Weighing heavily on my shoulders tonight is the weather forecast, which has predicted unbroken cloud; I'm seriously concerned about the event being a write-off if we can't see anything. But just as the final visitors arrive from outside on the viewing deck, I spot Richard poking his head out of the fire exit. I hear a quiet but definitely audible whoop. And then I know that things will be OK – the clouds must be parting. Our rain dances have worked.

The talk over, my fears dissipate entirely as I lead the groups in to use the shiny new telescopes. First up are the few children who have accompanied their parents; the other group members look on with smiling faces, but I know they want to get on the scopes as soon as possible. I switch on the red dark-sky light and the kids' eyes become charged with excitement as they scan everything around the room, trying to take it all in. It's a magical feeling. Their eyes rapidly settle on the 14-inch robotic telescope sitting proudly in the centre of the room on its large steel pier. Surprisingly for a group of kids, they wait for my instructions. I show a few of them how to flick on the power switch and the mount whirrs into life, releasing a cacophony of beeps and purrs to let us know that both axes are working. The room is full of giggling. 'It sounds like R2-D2,' one of the boys gleefully shouts. And he's right, it actually does sound like R2-D2.

Once the beeping stops, the scope is operational. But as the kids are quick to point out to me, the roof is closed. Before we see to that, we need to turn on my computer, which is tucked away on my desk in the corner of the room. The kids run over and boot it up; I don't need to ask them twice. We open the software and the screen lights up, silhouetting their little figures, which are huddled together in the twilight of the obser-vatory. We tell the program to 'Connect' and it does so, confirming its alignment with a further series of soft beeps. We are nearly ready for action.

I'm met by a sea of hands for my next request. 'Who wants to spin the turret?' On a cool night like tonight, spinning the steel handle and wheel by hand is a sure-fire way to keep warm, but it's also like a game for the kids. Each one of them takes it in turns to see how quickly they can spin the room around, and I find myself playing along too – I realise this contraption could become a highlight.

'What shall we look at then, kids?' I ask.

A pause.

'The Moon,' a voice shouts from the darkness.

'Yeah, yeah, the Man in the Moon,' someone else cries.

'Well, I'm sorry, the Man and his Moon aren't in the sky tonight . . . but the giant planet Jupiter is . . .'

'Yes! Yes!'

'Jupiter it is, then.'

I program the computer to find the largest planet in our solar system. I explain to them that Jupiter is a gas giant, a cloud-planet of hydrogen gas and fast winds that is over 300 times more massive than our Earth. In fact, we could fit over 1,300 Earths into its volume. This last bit blows their minds, as it does mine every time I think about it.

'Wait a minute,' I say.

They look towards me expectantly.

'Where is the sky?' The doors on the turret are still closed. 'We need to open the doors as soon as possible. Who can help me?'

There are four doors that form the opening in the turret, through which the telescopes will be aimed. The painted grey metal doors are held in place by electronically controlled metal arms at angles. These are controlled from a cupboard on the wall, which has a key to start the ignition. I ask a few children to walk over and turn the key; one by one the doors start to open. The process takes a few minutes, but it's worth every second for the gasps around me that are let out as the sky slowly starts to reveal itself.

'Look!' I shout, pointing to the deep shard of purple sky with fairy-light stars. 'Can you see that really bright one?'

The little heads look up into the night sky and one by one they shout back, confirming that they can.

'Well, that's not a star at all, it's a planet – that's Jupiter!'

The doors are now fully open; it's time for business. In goes the eyepiece and I check the focus, to ensure that it is aligned and ready. I ask a girl standing next to me to take the first look. She must be about nine. She climbs the small ladder to the telescope and takes her place in the hot-seat. It takes her about a minute to get used to looking into the scope; it can be quite a disorientating experience at first, knowing whether to look with both eyes or just one; the initial blur (before your eyes focus) can be confusing. But then she sees it: a large white ball of light, with two bands of dark material either side of its equator. She's delighted, and looks as proud as if she had been the first one to discover the planet, which in a way is true – she's discovered it for herself for the first time.

She likes the spot the most: the eye of Jupiter's storm, and asks her friends to come up and see if they can find it too. One by one, the kids climb up the small ladder and take a look: peering up at a planet over 600 million kilometres away, their little fingers clasp the knob and turn it towards focus. I can't feel any prouder. Their energy is such a jolt of adrenaline; I know that the mission for Kielder has been the right one.

I had originally envisaged the observatory as a mothership for all astronomers: the experienced, the armchair fans, as well as the just plain curious about what else is out there. It would become a beacon of exploration in a park of dark skies. And I knew it was all of those things. But I knew now that what I had really wanted to build all along was a home: a shelter for discovery where you could be safe and explore the skies, whoever you are. As I bade farewell to the final visitors that night at three in the morning, and shut up the gate, I knew this place would always be a refuge, a home away from home.

July–August Sky Guide

*Lyra – Cygnus – Aquila – Vulpecula, Sagitta
and Delphinus – The Perseids*

LYRA

In summer (August) looking overhead from Kielder Observatory

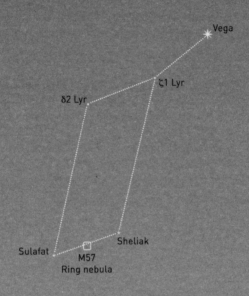

Vega

ζ1 Lyr

δ2 Lyr

Sulafat

M57
Ring nebula

Sheliak

Stars   Mag 0 ✸   Mag 1 ✸   Mag 2 ✴   Mag 3 ★   Mag 4 ·   Mag 5 ·   Clusters ☼   Nebulas ☐

Lyra is named after the famed instrument of Orpheus, the Greek musician whose lyre was so enchanting that it could charm even inanimate objects such as streams and trees. It's a special constellation to me and is the name of my Border collie pup, my gnawing companion as I wrote this book. Lying to the left of Hercules, Lyra is pretty and small – it ranks fifty-second by the area it covers in the list of all eighty-eight constellations. However, it is easy to spot thanks to Vega, its brightest star and the fifth brightest in the sky. If you have already looked at the glossary, you may recall that the magnitude system for stellar brightness is based on Vega, which is defined as having a magnitude of 0.0. Like Thuban in Draco, it sits on the precessing path of the North Celestial Pole and so acted as our Pole Star 14,000 years ago and will again in 12,000 years' time. It was also the first star after the Sun to be photographed. By the end of July you will find it due south at approximately 10.30 p.m.

Along with the stars Deneb (in Cygnus) and Altair (in Aquila), it forms a prominent seasonal asterism known as the Summer Triangle. Most depictions of the lyre have Vega at the top, connecting down to a parallelogram of four stars which constitute the main body of the instrument. First you'll encounter ζ^1 Lyr (a double), then moving around clockwise you'll come to

δ^2 Lyr (another double), Sulafat (γ Lyr) and Sheliak (β Lyr). Some sources suggest that the name Sheliak is derived from the Arabic for 'tortoise' in reference to the constellation's mythological origin. Far from the lyre itself, on the border with neighbouring Cygnus sits the famous variable star RR Lyrae. Astronomers use changeable stars like this to help measure distances in space.

Far and away the most famous deep-sky object in Lyra is the Ring Nebula (M57). Like many of the gas clouds we have seen already, this is a planetary nebula – the death throes of a star very much like the Sun. It was only the second planetary nebula to be discovered (after the Dumbbell Nebula in Vulpecula, discussed later). It is, however, a particularly striking example. You'll find it on the bottom edge of the lyre, approximately halfway between Sulafat and Sheliak.

As we are seeing it face-on, through amateur telescopes it will look as if someone has blown a small grey ring of smoke into the night sky. Through long-exposure photographs, however, its vivid colours are striking. As the layers of gas shed by the star expanded into space, they began to cool. So, closest to the centre the gas is still relatively hot and glows blue. But further out, you run the gamut of rainbow colours, ending up with much cooler red material. An iconic photograph from the

Hubble Space Telescope clearly shows these colours, as well as a star-like object at the centre. Known as a stellar remnant white dwarf, it is the leftover core of the dead star. About the same size as the Earth, it contains about half of the star's original mass.

Confident telescope users with fairly sizeable apertures may want to go in search of another Messier object – M56. This is a globular cluster, and sits approximately halfway along a line drawn between Sulafat and Albireo (β Cyg) in Cygnus. You'll be unable to resolve individual stars with binoculars, but you may still make it out as a slightly fuzzy star.

With a magnitude in excess of 13, a real challenge is the trio of colliding galaxies known as NGC 6745. Find it by starting with Sheliak, drawing the line up to δ^2 Lyr and continuing for about the same distance you have just travelled.

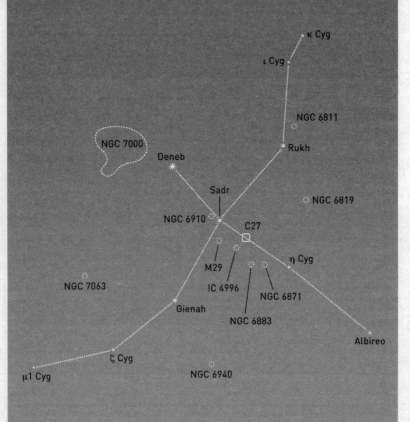

CYGNUS

In summer (August) looking overhead from Kielder Observatory

κ Cyg

ι Cyg

NGC 6811

Rukh

NGC 7000

Deneb

NGC 6819

Sadr

NGC 6910

C27

η Cyg

M29

IC 4996

NGC 6871

NGC 7063

NGC 6883

Gienah

Albireo

ζ Cyg

NGC 6940

μ1 Cyg

Stars   Mag 0 ✹   Mag 1 ✷   Mag 2 ★   Mag 3 ⋆   Mag 4 ·   Mag 5 ·   Clusters ⁘   Nebulas ▢

As constellations go, Cygnus is one of the more recognisable: the swan clearly has a long neck stretching through the band of the Milky Way, as well as a tail and wings that flap and fold. Its overall shape is often referred to as the Northern Cross – to match the much smaller southern hemisphere constellation of Crux, the Southern Cross.

Cygnus' brightest star – Deneb – represents the swan's tail. It also marks the top left corner of the Summer Triangle. Appearing at roughly the same brightness as the two other stars in the triangle – Vega and Altair – it is actually more than 100 times further away than Vega. That it can still match its neighbour for brightness serves to illustrate how inherently bright this supergiant star really is.

From Deneb, move into the Summer Triangle and you'll encounter Sadr (γ Cyg). The bird's wings extend out from here on either side through Gienah (ε Cyg), ζ Cyg and μ¹ Cyg on the left, and Rukh (δ Cyg), ι Cyg and κ Cyg on the right. But it's the star located at the beak of the swan that's the real spectacle. Called Albireo (β Cyg), it is widely referred to as the Jewel of the Summer Sky as it's the most beautiful double star in northern skies. Not only that, but the stars are separated by a considerable distance, so you don't need much magnification to prise them apart. A pair of decent binoculars should do the trick. What makes them so special is their beautifully contrasting colours – one electric blue and the other golden yellow.

There are many other, if less spectacular, double stars in this constellation: 61 Cygni can be found to the lower left of Deneb, slightly below the line between τ Cyg and ν Cyg. It was the first star to have its distance measured, via parallax, when it was calculated by Friedrich Wilhelm Bessel in 1838. The star at the left extremity of the wings – μ¹ Cyg – is also a pair, with magnitudes of 4.8 and 6.2 respectively.

Thanks to its position spanning the Milky Way, Cygnus is rich in deep-sky objects. At magnitude 5.5, the open cluster M39 is on the cusp of human eyesight and easily visible through binoculars. Find it by extending a line from Albireo through Deneb for about the same distance again towards the constellations of Lacerta and Cepheus. In the tail of the swan, just above Sadr on the way to Deneb, is the Rocking Horse cluster (NGC 6910). Visible as an indistinct grouping through binoculars, you'll need a magnification of at least fifty times in order to see why the cluster got its name.

Moving on to nebulae, perhaps the most famous in Cygnus is the North America Nebula. Covering an area four times greater than the Full Moon, you can see it just to the upper left of Deneb. It gets its name because its shape resembles the outline of the continent. Slightly closer to Deneb, almost between the stars 57 Cyg and 56 Cyg, is the Pelican Nebula. The two gas clouds are separated by a dark band of dust. In dark skies it is just about visible, as a brightening of the sky, but only just.

If you slip down the tail of the swan back to Sadr and head along the bird's right-hand wing, you will pass very close to the star θ Cyg. From here, jump up slightly to the double 16 Cygni. You're now right in the vicinity of the Blinking Eye Nebula. Its central star is so bright that through telescopes its light outshines the surrounding gas cloud. However, if you use averted vision and look out of the corner of your eye, you will be able to see the outer layers. Moving in and out of averted vision, the nebula appears to blink at you – hence the name.

The nebulae keep on coming in the form of the Veil Nebula, a 5,000-year-old supernova remnant. It is located just under the line between Gienah (ε Cyg) and ζ Cyg in the left-hand wing of the swan. Despite having an overall magnitude of 7.0, you'd think it would be relatively easy to spot. However, it has low surface brightness, which makes it a challenge, but try it anyway and enjoy it as the brighter region passes straight past the star 52 Cygni.

AQUILA

In summer (August) looking south from Kielder Observatory

ε Aql

Deneb el Okab

Tarazed

NGC 6709

Altair

Alshain

NGC 6755

Deneb Okab

η Aql

θ Aql

Al Thalimain Prior

Stars   Mag 0 ✳   Mag 1 ✦   Mag 2 ✶   Mag 3 ★   Mag 4 ·   Mag 5 ·   Clusters ⊙   Nebulas ☐

As mythology would have it, Aquila, an eagle, was the original thunderbird. Zeus would take his rage out on his enemies by using Aquila to deliver, and then retrieve, his vengeful thunderbolts.

The first magnitude Altair – Aquila's brightest star – forms the lower vertex of the Summer Triangle. One of the nearest stars to the Earth at just seventeen light years distant, it sits in the middle of the eagle's head, flanked on either side by Alshain (β Aql) and Tarazed (γ Aql). The bird's body then descends to the star Denebokab (δ Aql), from where its wings branch off towards Bezek (η Aql) and θ Aql on the left and ζ Aql and ε Aql on the right. Beneath Denebokab sits the tail star of λ Aql.

Bezek is particularly noteworthy as it is one of the brightest Cepheid variable stars in the sky. As their name suggests, these stars vary in brightness over time and in a very regular way. Bezek brightens and dims by almost an entire magnitude (between magnitude 3.5 and 4.4) over the course of about a week. With a difference of one magnitude equating to a real-terms brightness difference of 2.5 times, you should be able to make out this change for yourself.

Cepheid variables are particularly valuable to astronomers because they are used as one way to measure distances in space. There is a known link between the period over which their brightness varies and how intrinsically bright they are (their absolute magnitude). As starlight fades during its journey to Earth, we see all stars as dimmer than they really are. Astronomers can then tell how far away a Cepheid is from how much it has dimmed.

Around 5 degrees south-west of ζ Aql sits the open cluster NGC 6709. At magnitude 6.7, you will be able to spot it with binoculars, and with a small telescope you should be able to glimpse some of its forty individual stars. By contrast, the nebula known as B143-4 is an area devoid of stars because of its dark cloud blocking our view. At over a degree across, it will be conspicuous through binoculars because of that lack of stars. It is located 3 degrees north-west of Altair.

Those with a sizeable telescope and wanting a challenge should try to track down the planetary nebula NGC 6781, located less than halfway along the line between Denebokab and ζ Aql. Many amateur astronomers draw a comparison between this cloud and the Owl Nebula in Ursa Major. There is a globular cluster on offer here too – the magnitude-9.0 NGC 6760. On the way down from Denebokab towards λ Aql, locate the star 23 Aql just to the right of that line. Then continue on to the star 21 Aql. NGC 6760 sits below this pair to form an almost equilateral triangle.

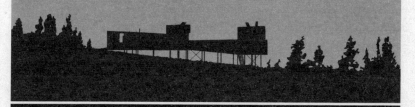

VULPECULA,
SAGITTA & DELPHINUS

In summer (August) looking south from Kielder Observatory

NGC 6885

15 Vul

VULPECULA

NGC 6834

Lukida

NGC 6830

M27

NGC 6823

η Sge   SAGITTA

γ Sge   δ Sge

M71   Sham

β Sge

DELPHINUS

γ2 Del   Sualocin

δ Del   Rotaneu

Deneb Dulfim

Stars   Mag 0 ✳   Mag 1 ✴   Mag 2 ✶   Mag 3 ✦   Mag 4 ·   Mag 5 ·   Clusters ⊙   Nebulas ▢

This trio of constellations may be tiny, but they pack a real punch in the terms of deep-sky objects. Unlike many of the other constellations we've seen, Vulpecula, the Little Fox, has little associated mythology. In fact, it was only formally identified in the seventeenth century. What's more, there are very few bright stars in the region. However, it is those deep-sky objects that make it worth seeking out. Located entirely within the Summer Triangle, it sits just beneath the star Albireo in Deneb. Its brightest star is Anser at magnitude 4.4 and is a double that is easily seen through binoculars. An even better binocular target is the Coathanger cluster (Brocchi's cluster). Start from Albireo and draw a line through Anser for a little more than the same distance again. You'll spy a line of stars with a hook appearing to extend out from the middle.

Head now in the direction of Cygnus's lower wing. After getting about a third of the way there, stop and search for the stars 12 Vul, 13 Vul and 14 Vul in the shape of an equilateral triangle. Just beneath 14 Vul sits Vulpecula's most famous deep-sky object: the Dumbbell Nebula (M27). Shining at magnitude 8.1, it was the first planetary nebula ever discovered. You can just about make it out with binoculars, and a telescope will reveal an hourglass shape.

It is worth a note to say that the Little Fox is also home to the first pulsar ever discovered. These tiny, dense and rapidly rotating objects – also known as neutron stars – are the remnants of supernova explosions at the end of a big star's life. Often only tens of kilometres across, they contain a considerable fraction of the original star's mass. Therefore their density is literally astronomically high. A single teaspoon of material from a neutron star would weigh as much as every person on Earth put together.

They are also highly magnetised and so beam radio waves out from their poles. We pick up these pulses with our radio telescopes – hence their name, which derives from a contraction of 'pulsating star'. These pulses are so regular that when English astronomers Jocelyn Bell and Antony Hewish heard the first one in 1967, they nicknamed it LGM-1 (or Little Green Men 1), joking that it could be from an alien civilisation. Today we know it as PSR B1919+21.

Between Vulpecula and Altair in Aquila lies the equally unremarkable constellation of Sagitta, the Arrow. It is the third smallest constellation in the sky, after Crux and Equuleus. Like Vulpecula, the mythology behind it is limited. Some stories describe it as the arrow that Hercules used to kill Aquila. Nevertheless, the head of the arrow is found at η Sge, with the shaft extending right through γ Sge and meeting the fletch at δ Sge. The arrow's vanes are marked by the stars Sham (α Sge) and β Sge.

The highlight of this constellation is M71, a very loosely packed globular cluster that at first glance looks like an open cluster. At magnitude 6.1, you should be able to spot it unaided from very dark skies, else a pair of binoculars will suffice.

For our final diminutive constellation we are heading just outside the Summer Triangle to the left of Altair. Here you'll find Delphinus, the Dolphin. The head of the animal is marked out by a squashed square known as Job's Coffin. It is formed from the constellation's four brightest stars, starting with α Del in the top right-hand corner. Moving around the quadrilateral anti-clockwise, you pass through β Del first, then on to δ Del before finally resting at γ Del. This last star is a very popular double star among amateurs. The dolphin's body then descends towards ε Del from β Del.

The naming of Delphinus' two brightest stars – Sualocin (α Del) and Rotanev (β Del) – is particularly notorious. As you may have appreciated by now, astronomers have very strict naming conventions. In particular, it is forbidden to name stars after yourself. In fact, there are very few stars in the night sky named after people at all, and in those cases they were named in their honour. Now, no one is called Sualocin or Rotanev. However, if you look at those names in a mirror, they suddenly transform into Nicolaus Venator, the Latinised form of Niccolò Cacciatore, a director of the Palermo Observatory in Sicily, Italy. He published the names in a star catalogue of 1814, but it took forty-five years for anyone to notice his ruse. By then it was too late, and so he remains the only person to have put his own name to a star.

As well as a good back-story, Delphinus also has a beautiful deep-sky object to offer. NGC 6934 is an 8.8 magnitude globular cluster which can be found beyond ε Del at the tail of the dolphin.

The Perseids

August is home to one of the most celebrated events in the astronomical calendar: the Perseid meteor shower. Technically the shower lasts for over five weeks, starting to build in mid-July before petering out by the end of August. However, it reaches a spectacular peak around 12 August every year when shooting stars appear to rain down from the sky at the rate of at least one a minute.

Except that they are not stars at all. Instead, these meteors are tiny grains of comet dust burning up in the Earth's atmosphere. Many comets – ancient cosmic ice mountains – patrol the solar system, with the most famous example being Halley's comet. Although it had been seen many times before – it even appears in the Bayeux Tapestry – its return was predicted by English astronomer Edmund Halley, after whom it was named. Unfortunately he didn't live to see his prediction come true.

When a comet's path brings it close to the Sun, our star's intense heat vaporises some of the ice, which liberates dust. So, Hansel and Gretel-like, comets leave a trail of crumbs scattered throughout the solar system. When the Earth's orbit then takes us into that trail, the crumbs meet a fiery end. The comet responsible for the Perseids is Swift–Tuttle. With the comet only orbiting the Sun once every 133 years, the incinerated dust we see has built up over many millions of years.

Meteor showers get their names from the constellations they appear to radiate from, in this case Perseus (see September–October guide). However, for the best view you shouldn't stare directly at the radiant constellation as this is where the meteors are coming from, not where they are going to. Before midnight on the days around the shower's peak, Perseus sits low near the north-eastern horizon. Cast your eyes around in this quadrant of the sky and you're sure to see them. The constellation of Cassiopeia can act as a very good guide for where to look.

As with all astronomy, it pays to get away from sources of light pollution both natural and man-made. In urban areas, the intense glare of street lighting will wash out all but the brightest meteors. You might only see one every five or ten minutes. The Moon, too, can be a hindrance. Its bright light can have a similar effect to street lights, so try to wait until it sets to go meteor-watching. Whilst you certainly don't need binoculars to see the brightest meteors, they can definitely help you catch some of the fainter ones. Setting up a DSLR camera on a tripod and angling it towards this region can result in spectacular photographs too.

Whilst the Perseids are the most lauded, they are far from the only annual meteor shower. Here are some of the most notable alternatives:

Name	Peak	Maximum number of meteors per hour
Quadrantids	3 January	120
Lyrids	22 April	18
Orionids	21 October	20
Leonids	17 November	15
Geminids	14 December	120

I'm often asked what the difference is between a meteor and a meteorite. It all depends on the location of the object you are talking about. A small piece of debris orbiting the Sun – perhaps a broken-off piece of an asteroid or comet – is known as a meteoroid. They are less than one kilometre across and often considerably smaller than this. If that meteoroid enters Earth's atmosphere, then friction with the layers of gas causes it to heat up and we see an intense flash of light colloquially known as a 'shooting star'. It is this light that we call a meteor. Only if a meteoroid survives the treacherous journey and makes it to the ground is it called a meteorite. Around 40,000 tonnes of material is added to the Earth from space each year. So a meteor is a 'shooting star' and a meteorite is one that's landed on the Earth and you can pick it up.

The world's collection of meteorites comes from three main sources: the asteroid belt, the Moon and, in some extreme cases, Mars. In these latter two examples, collisions with those bodies have broken pieces off, some of which have made it across to the Earth. Those that come from the asteroid belt are pristine relics from a time before the Earth and the other planets existed and so help scientists piece together the history of our solar system.

If you look at the Moon through binoculars or a telescope, you will see many craters, evidence of impacts from asteroids and comets. Look around the Earth, however, and we don't seem to have as many. So you might think the Moon has been hit more than we have. But a bigger body with a stronger gravitational pull, the Earth has actually been hit more. We just have an atmosphere to prevent smaller impacts reaching the surface, and water, weather, earthquakes and volcanoes to cover up much of the evidence if they do make it to the ground.

Another question we frequently get asked at the observatory relates to comets and why they have tails. Actually, most comets have two tails, both of which can stretch for many millions of kilometres. They begin to form as the comet approaches perihelion, the point where it is closest to the Sun. Here the Sun's energy starts to vaporise some of the comet's ice, releasing dust into space. The first tail, known as the 'dust tail', is formed when pressure from sunlight pushes these dust particles away from the comet. As the pressure associated with sunlight is relatively tiny, these tails are usually very diffuse and are often curved.

The second tail, known as the 'ion tail', is caused by the Sun's ultraviolet light. It has higher energy than visible light and so is able to strip charged particles from the comet. These particles then interact with the charged particles in the Sun's solar wind and are deflected directly behind the comet in a straighter, narrower tail than its dusty counterpart.

In rare cases, a comet can have a faint third tail made of sodium. A famous example is the comet Hale–Bopp, which became one of the brightest comets to blaze through the sky in the twentieth century when it was observed in the mid-1990s.

5

Space Travel

Here the land and sky meet. Red sandy hills bulge towards the cloudless blue at an altitude of over 7,000 feet. There are no trees, no scrub – no people for that matter. Nothing lives here; it is too dry. Windswept dunes and odd rocks resemble the surface of Mars, or Luke Skywalker's desert planet of Tatooine. It's not at all like Kielder, with its evergreen forests and its sheep. I can't help thinking these plains haven't changed much in a million years.

In the desert air my throat is dry. I reach down to my side for a bottle of water as Glenn slips heavily out of the passenger door.

'I'm . . . bloody speechless,' Glenn manages to sputter out.

I resist calling him out on his 'speechlessness'; it's been a very long trip.

'Me too,' I reply, slowly looking around with a smile widening on my face, breathing it in.

We share a tired laugh and stumble to the back of the camper van to open the rear door. Out drops Adam. For the next few minutes the three of us stand shoulder to shoulder in silence, staring at the desert and the endless sky. We are the only ones here. We can't work out what is more profound: the vastness or the absolute quiet.

It is 14 May 2013 and we have been planning this trip for over a year: it is a pilgrimage I have wanted to make for most of my adult life. After the long-haul flight, we drive for seven hours through some of the most spectacular valleys in the world, surrounded by mountain ranges and active snow-topped volcanoes. There are no real towns or cities of note that we pass along the way – just long empty roads dissecting the desert with their straight lines, and the odd large truck rumbling through the sands. But despite the barren plains, I can feel this is an unmistakably special place: surrounded and wrapped up by the biggest of aquamarine ceilings.

We eventually find our spot for the evening, a dusty track leading into a small nature reserve. We park up our beaten camper for the night, a three-berth van complete with shower, cooker and fridge. As night closes in around us, I get a shiver of excitement. I have brought quite a bit of kit with me, including a small imaging set-up that consists of a tripod and several cameras, as well as a telescope to search for faint distant light emitted by objects nestling in these darkest of skies. Transporting masses of equipment over here has been a feat in itself, raising a few eyebrows at check-in. But reading out the stickers on the boxes to the baggage staff was a proud moment: 'Astronomical instruments: Handle with Care'. It felt like we were on an important mission, I felt like a scientist.

Adam is the chef and sommelier among us and is eager to open his latest find, a bottle of Argentinian Malbec. He is here for other reasons, of course, but he excels at boozy desert BBQ. I have different ideas. I know I only have around ninety minutes of sunlight left before it will begin, so I set about assembling my kit quickly. I'm a little nervous at first that it might have become damaged in transit or, still worse, that I've forgotten something important, but thankfully out it all comes, in one piece. I gently reach for my prized possessions first: my beloved Canon 60Da and its older brother, the 20Da, both well used but spotless for the trip; I can still smell the lens cleaner. Next come the chargers and the cables, a spool of tangled mess. Finally I pull out the AstroTrac mounting system, which will guide the cameras as the Earth rotates. Setting up the legs on the ground, I realise that I love this ritual. Even after doing it a hundred times, when I stand back and see all my kit gleaming and raring to go, there aren't many better feelings. Even better still when I know we have an excellent forecast for the night. The sky is now a blue-pink as the sun begins to sink down. After a quick twilight dinner of roast chicken over an open fire, washed down with a cold beer aperitif, I power up my laptop. The telescope and cameras bleep pleasingly in response.

We sit back and wait. And wait. Twilight first offers us up Antares in the constellation of Scorpius, a behemoth of a star pumping its yellow light towards Earth. Scorpius, the scorpion, curls its way through the centre of our galaxy, with an arc of stars signifying the celestial sting in the tail. It is a huge winding constellation, with its knots of dust and gas intermingled with the stars. The background sky gradually darkens and we begin to see what we have come for. I know it well – I have read about it and studied it at length. I have observed it many times from Kielder, but never like this. Not from skies that are so dark and generous the scientific astronomical community from around the world travel here. 'Can you see it?' I say to the lads. I point my laser pen to the sky, shooting up a solid green beam to trace the outline of the celestial object. Not yet.

Night gradually takes hold and the sky descends into inky black. About an hour later it comes sharply into focus. Glenn and Adam can see the object clearly without my pen; it fills the whole sky.

'What on earth is it?' Adam whispers quietly. Any louder and we might scare it away. Glenn is speechless again; it is beautiful beyond words. I look at them both.

'It's the Galactic Center, the Milky Way, our home galaxy.'

*

To witness the Milky Way in all of its glory you need the darkest of skies, and there is no better place on Earth to see the bounds of our galaxy than the Atacama Desert in Chile. There are many sites throughout the world where darkness reigns supreme, free from the scourge of light pollution. Kielder is one of the prime locations in the UK, but desert regions at altitude top the list. Their low levels of humidity, combined with the thinner atmosphere, offer up some of the most crystal-clear panoramas of the cosmos on the planet.

Chile has the darkest and driest skies on Earth, which make it the natural home for some of the world's biggest and most advanced optical telescopes. Most are found at the Observatorio de Paranal, built on one of the highest peaks in the Atacama Desert. Here, astronomers, scientists and researchers gather to use its famous set of four telescopes, collectively known as the VLT, an abbreviation for the Very Large Telescope. Situated above the inversion layer where clouds form, at over 7,000 feet above sea level, and with only 5 to 15 per cent humidity, the observatory theoretically offers astronomers over 350 clear nights every year.

It is the day after our desert BBQ, and as the road flattens out and the sun begins to set, I know we are

A land pier, or a spaceship . . . The observatory took over a year to build, constructed by a team of expert carpenters. The wind turbine was erected to help power the telescope's computers along with the solar panels and the generator. The poet Alec Finlay composed the words on the blades of the turbine, which together form a circular poem: 'space arcs, light eclipses, time bends.'

Finished at last . . . The Kielder Observatory was officially opened in April 2008 by the 14th Astronomer Royal, Sir Arnold Wolfendale from Durham University, a mentor and now a friend.

At night the observatory comes to life . . . My colleague Dan Monk, one of our youngest members of staff, peers up into the cosmos.

Star trails . . . if you point your camera north in a long-exposure photo, the stars will be seen to track around the north celestial pole.

Below: The Milky Way arching overhead at Kielder.

The aurora borealis, also known as the Northern Lights, can be seen from Kielder on a clear night when there is plenty of solar activity.

Space travel . . . As part of my public outreach work for Kielder, I have been fortunate enough to travel to Paranal Observatory, located in the Atacama desert in Chile. It is the home to some of the darkest skies on earth, as well as some of the world's largest, most advanced telescopes. The silhouetted picture above shows me staring up at the VLT (Very Large Telescope).

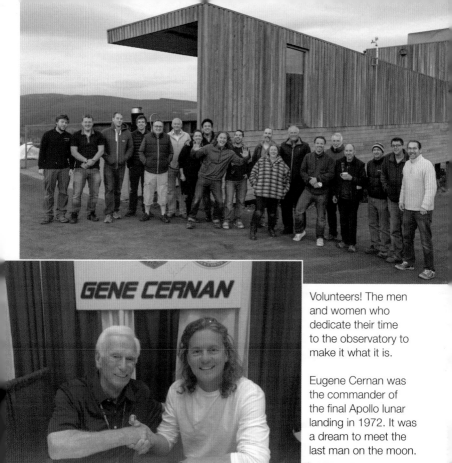

Volunteers! The men and women who dedicate their time to the observatory to make it what it is.

Eugene Cernan was the commander of the final Apollo lunar landing in 1972. It was a dream to meet the last man on the moon.

As it was to visit the Hooker telescope at the Mount Wilson Observatory in California. The Hooker was used by Edwin Hubble to prove that the universe goes beyond the Milky Way galaxy and is expanding.

At home in Kielder.

Sky patrol . . . clear forecast.

Accepting an honorary Masters in
Science from Durham University at a
ceremony at Durham Cathedral.

A handful of stars . . . my dog Lyra,
named after the constellation.

close: the observatory I have dreamed about for years is now just a few minutes away. 'Dip headlights during darkness' the sign reads, as we make the final approach. The van's tyres spin in the dust and gravel, and my stomach lurches with anticipation. After a two-hour drive from our campsite the night before, we take one more meandering left, one last climb. I can't help shouting out loud. I cut out the transmission and yank on the hand-brake. I have to get out.

Standing alone in the middle of the road, the sun warm and the wind blowing steadily on my sunburnt cheeks and lips, I look ahead. In front of me, four cylindrical steel turrets rise up, shimmering in the heat on the table-top mountain, my favourite observatory in the world. Resembling a rocket launch site in its futuristic mini-malism, only the telescope housings and a few satellites populate the desert peak, which is surrounded by barren hills and sand dunes. The real attraction is up above. I throw my head back with a laugh. I'm staring up at the deepest blue I have ever seen. I can't believe my luck.

My golden ticket has come about through a chance encounter with an acquaintance I met at Durham University a few years ago. Dr David Murphy is a post-doc working here in Chile, but today he is like a super fixer: he has granted us a five-star, behind-the-scenes pass to visit all of the country's biggest observatories, including the VLT at Paranal. Glenn and Adam have accompanied me on the trip to shoot a documentary we are working on, entitled *Searching for Light*. Well, we have nearly found it.

We are waved through the gates by a friendly security guard, a Chilean local; but his welcoming smile belies the heavy surveillance that monitors our every move; this compound houses some of the most expensive and sophisticated instruments in the world. To add to the James Bond allure, tonight I will work with a Frenchman named Julien Girard, but fortunately he's no master villain; he's a generous young astronomer who kindly shows me to his office. On cue, as cool as you like, the aroma of freshly brewed coffee wafts over from his desk. He explains that the night shifts at the observatory are long, often twelve hours, and that caffeine is essential. We quickly fall into conversation and I bombard him with questions. I'm surprised when he tells me that scientists like him don't actually operate the scopes at Paranal; rather, that is a job for the technicians. Instead, he is in charge of the imaging and evaluating the data. On Julien's desk there are three computer screens: one for the scope and drives; one for locating the objects; and one for recording the data or monitoring weather conditions such as humidity and visibility levels.

Before it is dark and the science starts, however, Julien asks me if I would like to go outside onto the deck with him to join the rest of the office. Sure, I say, why not? I know the sky is getting darker and I'm eager to see it, but I'm a little puzzled by the request as night hasn't fallen yet. I'm still confused until we venture out onto the deck where the other scientists have gathered. I let out a short gasp. What office in the world evacuates completely to go out and see the setting sun?

I don't know what to say or where to look: it is a spiritual experience. We are standing in the heavens; the cloud tops beneath us a golden yellow and the sky glows pink through lavender and deep blue. A bright orange/red orb slowly drifts south to meet the clouds. Silence descends as we wait, watching almost in disbelief. I am reminded again of nature's awesome power: ours to share, never to control or own – just to look at, to feel a part of.

When the sun finally sets, there is a brief moment of incandescent refraction – an earthly starburst. Then the colours gradually mute and darken, and one by one the scientists disappear into the control room to recommence their shifts. I stay a little while longer, watching Venus chasing behind the Sun.

I'm interrupted by the sound of motors. The huge telescopes are blinking, opening up their shutters to disclose their 8.2-metre-wide eyes to the sky, ready to capture photons of light – photons that have potentially been travelling for billions of years through the cosmos from the most distant light sources. It is time to go inside and Julien pats me on my shoulder. I follow him with Glenn and Adam into the control room, where we will spend the next few hours learning and gathering fresh data about the Milky Way.

*

Minus a few faint, fuzzy galaxies, every star, every point of light that you have ever and will ever see in your night-time sky is inside the Milky Way, the galaxy in which we live, home to over 300 billion stars, many of which may have their own retinue of planets.

On a clear moonless night from a dark-sky location, such as Kielder, you can see the Milky Way as a band of bright light that stretches across the sky. It is faint but laced with brighter regions that lie among its arced structure. I said at the beginning of this book that the shape of the Milky Way resembles a Frisbee: if you look at it front-on (from outside our galaxy, which we cannot), it appears circular, but if you look at it side-on through the plane of a disc (or from inside it, as we do on Earth), it resembles a straight line that bulges slightly in the centre (like the profile of a Frisbee).

The stars we see in the night sky that appear to be situated outside his Frisbee-shaped structure are actually still within it, but just much closer to us. Confusing as this seems, if you were to zoom away from the Milky Way, what you would see would be the band of stars slowly condensing into one structure.

That evening with Julien at Paranal, before we went outside to see the Milky Way with our own eyes, was spent discussing research into the centre of the Milky Way. He explained that when we look towards the bulging centre (through the plane of its disc), our view is obscured by a collection of giant molecular clouds of dust. It is like looking towards Earth's horizon on a sunny day: the closer we look to the horizon, the more clouds we see; but it's not that the sky is cloudy, it's more to do with the fact that we are looking through the plane of our atmosphere, which makes the clouds that are quite distant appear closer and more condensed, so blurring our vision. In the same way in the Milky Way, interstellar clouds of dust obscure the stars gathered at the very centre of our galaxy, so optically we cannot see them. However, one of the properties of dust clouds in space is that as starlight shines through them, visible light is absorbed and then re-emitted in infrared wavelengths. An infrared camera can therefore detect the location of these stars that were previously hidden from view. Using an infrared camera on one of the giant telescopes in Chile, scientists like Julien have made some of the most important scientific discoveries of the last few decades.

In a high-tech game of detective, the team at Paranal have observed, over a period of eleven years, the motions of stars around the centre of our galaxy in a region known as Sagittarius A*, in particular a star named S2. They noticed that S2 would speed up significantly on its orbit as it approached another object, which it appeared to be spinning around. In fact, it would speed up so quickly that it would briefly reach close to 5,000km per second, a significant proportion of the speed of light. In order for a star to whizz around a gravitational point this quickly, although they couldn't see it, the scientists knew the object would have to be massive.

Using gravitational laws, the team was able to monitor other stars too, and by comparing each of the orbits individually, they could determine that the object around which these stars were spinning, which emitted no light, had to have a mass of 4.26 million solar masses. That is, it was 4.26 million times the mass of our sun.

The scientists had discovered a supermassive black hole, one that was so massive that its gravity was curving space to the extent that S2 was speeding around it but never falling in, because it was travelling too fast for that to happen. The irony wasn't lost on me that in one of the darkest places in the world, where the clearest views with the naked eye can be seen, the greatest scientific advances being made are with objects that can't even be seen with optics. We stayed with Julien a little while longer, but soon I had to go outside to see the Milky Way for myself – the old-fashioned way.

The Milky Way is an elusive structure to observe, but surprisingly, even in the West, there are many places to see it, as long as you have a dark sky. However, from Kielder or the Atacama Desert in Chile, the Milky Way lights up the sky and is a sight to behold. Walking back outside to the deck of Paranal, my mind frazzled after hours of heady conversation, I came to my senses quickly in the cool air. Arched gloriously overhead, spanning horizon to horizon, was an unimpeded view of the Milky Way. Stretching out and streaming through many constellations, its star fields were rich with dark knots of nebulosity that snaked and weaved through the sky. Star clusters glistened like diamonds; still brighter stars with obscure names like Antares and Sirius burnt yellow and piercing blue-white. The Milky Way got brighter and wider as I looked to the southern part of the sky, and the galactic centre, as it is known – home to the supermassive black hole on which Julien's colleagues were working that night – pulled me in. I felt as if I could actually reach up and touch the Milky Way; it felt three-dimensional, as of course it is. The Milky Way seen from Kielder is impressive; the big difference here was that it was directly overhead, and because of the low humidity it was darker and clearer than I had ever seen it before. It was as if a fog had lifted, and I was seeing someone again for the first time.

I whisked out my scope, laptop and camera from the van and set things up as quickly as I could, barely able to contain myself as I took my first picture, a two-minute exposure that slowly downloaded to my computer. When the image was eventually displayed on the screen, I was shocked. Star fields littered with clusters, dark lanes of dust meandering through the billions of distant suns, whose light had been travelling for thousands of years. I could see the image interlaced with glowing pink regions, the tell-tale sign of distant, star-forming gas clouds, stellar nurseries where the next generations of new stars are forming. It was one of the best images I had ever seen, let alone taken. Glenn and Adam looked suitably impressed.

That night, looking into the heart of our Milky Way galaxy from Chile, took me back to the first time I had witnessed the same stars as a child. I remember that night well, too. I was eight years old in a holiday park in Devon, southern England. My parents had taken us there with my cousins, my aunt and uncle, and of course my two brothers. I would so look forward to those holidays: days filled with beaches and sunshine, sunburnt shoulders drenched in margarine (no sun cream then). We were all together, which is why it was special. It was the end to another balmy day and we were still running and playing outside in the darkness, my dad joining in as we panted for breath and were chased one more time. I remember the blackness; I remember stopping dead in my tracks and looking up. I have no idea what gripped me or why I suddenly looked skywards, but I did. I could see brightness all around, stars shining. My dad later recalled telling me, 'Gary, what can you see, what are you looking at?' Apparently I pointed to the sky and asked, 'What's that, Dad?'

He showed me the arc of the Milky Way. Back then

there would have been even less in the way of light pollution in the UK, so the skies would have been darker and there would have been even more stars to see with my young eyes. Now, looking skywards in Chile, I felt like a child again, grabbing my dad's hand. Except now I was the guide. He would have got such a kick out of standing with me under the world's most advanced, multibillion-dollar telescopes where the greatest discoveries in astronomy are being made. You never know how far life's journey will take you.

*

Thanks to astronomy, I have had the privilege of travelling to some of our planet's most beautiful far-flung corners, to meet some of the most brilliant minds in science, all in the search for light. Before I journeyed to Chile in 2013, that year I also visited the Arizona Desert and the Tenerife Caldera. The following year I was lucky enough to travel to Pasadena, California, for Spacefest, where some of the most important figures in the history of space exploration had gathered. After each of these trips, I returned to Kielder feeling a bit wiser and a bit more inspired, and I've tried to share these stories and experiences with our visitors.

I have met ordinary people who have achieved extraordinary things, such as Thomas Bopp, an amateur astronomer who was working in a construction factory in Arizona until one night in 1995 he discovered his first comet, which just happened to be the brightest and most dramatic comet that has been observed in living memory. Hale–Bopp, as it is now known (after Thomas Bopp and Alan Hale), blazed a dazzling trail across the sky in the northern hemisphere for nearly eighteen months, a record for naked-eye observing that doubled the previous record set nearly 200 years before by the Great Comet of 1811. What is even more remarkable is that Bopp didn't even own a telescope at the time. He was outside with a group of friends one evening, observing star clusters in Stanfield, Arizona, when he discovered a bright object as he looked through a friend's 17-inch scope, an instrument no bigger than the ones we have at Kielder. It dawned on him that he might have found something new when he checked his star charts and noticed that there weren't supposed to be any known astronomical objects in the sky near M70. He contacted the Central Bureau for Astronomical Telegrams, who confirmed his hunch, and the rest is, well, history.

I have also met people who have pursued fields of study most of us would consider beyond the realm of mortal man, even against the odds of debilitating illness. I was introduced to one of my heroes at a global astronomy event called Starmus in Tenerife. I had been invited to spread the word about the Kielder Observatory, and I walked bug-eyed into the foyer of the grand hotel. Before I locked eyes on my hero, I saw someone else first out of the corner of my eye. It was the hair. I mean, I have enough of the stuff myself, but in my peripheral vision those famous unruly curls were unmistakable, now a shade of silver compared to his jet-black rock and roll days. Wearing an over-sized bright Hawaiian shirt, it was the organiser of the event, Brian May, the former Queen guitarist and now a serious astronomer. He warmly said hello. Who said stargazing isn't a broad church?

It was the following evening, however, when I wandered downstairs from my room to the fancy dinner, with a suitably astronomical cost, that I first saw in the flesh the man whose work I have followed for much of my adult life. I have read his books many times and still do, revisiting them often to gain a better understanding, to make sense of a concept or an equation. He is a man I hold in as much regard as Einstein, and like Einstein he doesn't take himself too seriously – not enough theoretical physicists muse on the probability of universes existing where the moon is made of cheese. I adore the singular sense of humour in his writing. But now, off the page, there he was, sitting across the table from me in his famous chair, surrounded by his team. Of course, at first I didn't have the courage to just wander over and introduce myself. But after a few glasses of wine, I finally steeled myself. I spoke briefly with his nurse, who kindly introduced us.

There was so much I wanted to say to Professor

Stephen Hawking, but couldn't. He has inspired so many people, not just with his work, but in his long fight with motor neurone disease and his undimmed thirst for science and life. I mumbled a few platitudes and explained what we were trying to do for public star-gazing in the UK at Kielder. He looked at me silently and smiled, he didn't say a word. But he seemed to say so much. His stare was deep and behind those eyes I knew a mind of immense capability was whirring away. I felt in awe of him. After a few minutes I had to bid him good night and walk away. But for a moment, I would like to think that as our watery eyes made contact, he knew that I, like him, was alive for one passion – the universe – and the pursuit of understanding it. Or so I told myself.

If I organised a dream astronomer's dinner party, Professor Stephen Hawking would be at the head of the table. Sitting at the other end could well be the late great Sir Patrick Moore. If it hadn't been for his unforgettable voice and monocle, and his wonderful appearances on *The Sky at Night*, which kept me in touch with astronomy during my building years, I'm sure that I and a generation of others would have floundered. As a small homage at Kielder to Sir Patrick after his death in 2012, we were honoured to rename the large turret that houses our biggest telescope 'The Patrick Moore Observatory'. His influence cannot be understated.

However, if I were to pick the most memorable

encounter during my space travels, it would have to be during my meeting with one of last century's greatest space aviators, on a trip to a historic site where our knowledge about the universe really began to unspool. It certainly had the largest impact on me, and on our mission at Kielder today.

*

15 March 1929. From the Mount Wilson Observatory in Pasadena, California, Edwin Hubble makes a startling discovery that will rock the scientific world. Using the 100-inch Hooker telescope, he posits a theory that will cause fierce debate for the next hundred years. Our universe is expanding.

Fast forward to 13 May 2014, and I am once again driving up a twisting road, this time flanked on either side by pine trees scorched dry in the heat; I swerve to avoid the cracks in the tarmac split by the searing heat of solar radiation. In the shadow of the mountain, stretching into the distance is Los Angeles, the sprawling metropolis where human stars are churned out in their thousands. From my vantage point, I'm reminded of a line from a Neil Young track about junkies and setting suns. I like it up here.

Glenn is with me at Mount Wilson Observatory to film another short documentary. Although the tele-scopes here still provide stargazing sessions for the public, and some research is still conducted, Hubble's discovery was its greatest. The observatory's halcyon days of cutting-edge science still remain; however it is now mainly a museum. A sprightly janitor welcomes us inside and escorts us up a staircase with whitewashed walls. I'm struck by how clean the place is; immaculate is the only word to describe the observatory's brilliance. We round a corner and there it is. I recognise it from the famous black-and-white photograph: Albert Einstein peering into the elaborate lens; Edwin Hubble looking on and puffing into his pipe. Up close, the enormous blue-painted telescope is beautiful, carrying an arsenal of knobs and switches, all antiquated and resembling something from a Jules Verne novel. Beneath the tele-scope is a rickety wooden chair with a high arched back which Hubble would sit at for hours, turning the gears and motors by hand with the help of his loyal assistant, Milton Humason. I long to sit in the chair; to know what it must have felt like to be here in this room, gazing through the lens when all of these discoveries were made. Hubble has been described by Lawrence Krauss, a contemporary theoretical physicist, as a beacon for humanity. He started off in life as a lawyer before he became an astronomer. I wonder what he would have made of my trajectory . . . As my mind trips and races, the janitor unfortunately catches my eye and points to the sign: no sitting in the chair.

Half an hour later, my phone starts to ring and it is a contact at the convention centre. I pick up. 'Gary, you've got Cernan! He wants to talk to you. Get down here now, as fast as you can.' I turn to Glenn, who has heard every word, and he scrambles to get his bag. I think it's the fastest I've ever seen him dismantle a camera rig. We thank the janitor, pay our respects to Hubble's chair and dash for the Dodge. Half an hour later I'm sitting across a desk from one of the twelve people in history who have walked on the Moon.

'Dream about the impossible – and then go make it happen.' The words of Eugene A. Cernan, better known as the last man to walk on the Moon. On 7 December 1972, just forty-one months after Neil Armstrong first took his giant leap for mankind, *Apollo 17* took off from the Kennedy Space Center in Florida, commanded by Gene Cernan, a former naval officer and fighter pilot. It was the final flight of the *Saturn V* rocket, the glorious chariot of the Apollo programme, and the final lunar landing.

Cernan is sitting opposite me at Spacefest, one of the largest gatherings of space enthusiasts in the world. Wearing a crisp blue NASA shirt, he sits bolt upright; tall with a strong head of silver hair, his arms are ropey and powerful. He is a proud man and still commands respect; you would never think he was eighty years old. Cernan flew into space three times: first as pilot of *Gemini 9A* in 1966; then as lunar module pilot of *Apollo 10* in 1969; and finally as Commander of *Apollo 17* in 1972. On his final trip he spent three days living on the surface of the Moon. He drove more than 35 kilometres using the Lunar Roving Vehicle and spent much of his time with colleague Harrison Schmitt, the first geologist on the lunar surface. Together they collected rock samples and explored uncharted territory. One of their lesser-known triumphs is that together they hold the official land/lunar speed record – an impressive 18 kilometres per hour was reached with the use of the lunar rover.

Cernan doesn't need much from me in the way of questions. He is prepared and has a message so well rehearsed that it is almost political, which is not too much of a stretch – in 2010 he testified with Neil Armstrong in front of the US Congress in opposition to President Barack Obama's cancellation of space-flight plans to return humans to the Moon or to fly to Mars. He is passionate about this subject; young boys and girls should have the chance to go where he went, to explore and go even further afield.

He stares intensely at me and continues. 'The essence of human beings is curiosity, we want to know; we just gotta know,' Cernan says as he begins his final monologue. 'When I was up there, in the heavens as we call it, it was like I was sitting on God's front porch. And when I was leaving the Moon, as I was looking down at my final footsteps, I knew I wasn't going to be coming that way ever again. Someone would, but it wouldn't be me. And I was trying to figure out what we, not just

Apollo 17, but what we as a whole generation actually accomplished with space flight during those years. What was the meaning of what we did? We furthered technology, sure. But during those last steps, when I looked over my shoulder and looked back at the Earth, at our identity with reality, at the real world we live in, I realised that all life and love and family was all back there. I wanted to come back to Earth to be able to tell you, and whoever might be listening, particularly the younger people, what it was like. I wanted to reach out and grab that beautiful Earth and tuck it in my spacesuit and say, "Hey, here's what it looks like, here's what it feels like." That wasn't possible, so I came home with somewhat of an empty feeling, because you can't answer those questions with a phrase or two. It's an emotional experience that you have to find a way to relate to people.'

Our interview came to a close. With his final words, Cernan said that, decades later, he finally felt as if he was accomplishing his goal. He was going to spend his remaining time communicating the extraordinary potential of space, and his own experiences, to people around the world. It dawned on me that the Apollo legacy is a vitally important message; that we can achieve great things if we dare to question and try. Whilst most people may be interested in the first people who walked on the Moon – Armstrong and Aldrin – the last man on the Moon is more profound for me, and far more inspiring. Maybe Cernan's own personal history is being lost. In this age of supercomputers and high-tech space flight, maybe the stories of the pioneers of space flight are being forgotten. But his thoughts resonated with me.

I left Pasadena with a new mission: to share the stories of these adventurers, of how daring and exciting they were, with new generations back at Kielder. It was through the bravery and dedication of these few that we as a race have reached out into the cosmos. Space travel is crucial to our imaginations, and if anyone wants it badly enough, they have the opportunity. It all begins with looking up, and dreaming of being amongst the stars.

September–October
Sky Guide

Perseus – Cassiopeia –
Andromeda – Cepheus – Jupiter

PERSEUS

In autumn (September) looking south from Kielder Observatory

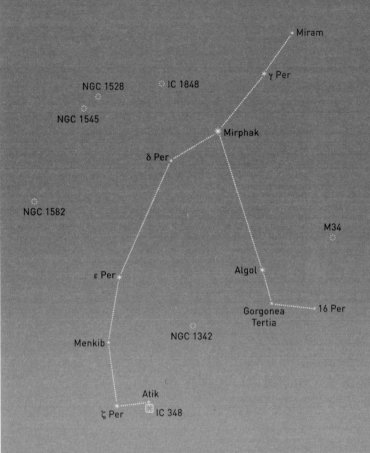

Miram

γ Per

NGC 1528

IC 1848

Mirphak

δ Per

NGC 1545

NGC 1582

M34

ε Per

Algol

Gorgonea
Tertia

16 Per

Menkib

NGC 1342

Atik

ζ Per   IC 348

Stars   Mag 0 ✺   Mag 1 ✳   Mag 2 ✶   Mag 3 ✦   Mag 4 ·   Mag 5 ·   Clusters ⊙   Nebulas ▢

Along with Hercules, Perseus is one of the most famous heroes in Graeco-Roman mythology. The rider of the winged horse Pegasus, Perseus was the one to cut off the head of Medusa, the Gorgon who could turn anyone to stone with just a look.

As the Sun sets at the end of September, you'll find Perseus just beginning to rise in the north-east. The brightest star in the constellation is Mirphak, which sits at the hero's navel. Located smack bang in the middle of the band of the arching Milky Way, it is surrounded by many dimmer stars, making the region a treasured target with binoculars.

Yet it is Perseus' second brightest star – Algol – that is arguably the more famous. Known as the Demon Star, it represents an eye in the severed head of Medusa, which Perseus is holding aloft in his right hand. It is known in Arabic as *Ra's al-Ghul* (a name fans of *Batman* may well recognise). Algol is one of the most notable variable stars in the sky. Over the course of around 2.9 days it will brighten from magnitude 3.5 to 2.3. This is because it isn't a solitary star but a triple star system whose components are constantly eclipsing one another and therefore changing the overall amount of light we receive.

To complete the rest of Perseus, first start with Mirphak and climb his torso through the magnitude-2.9 γ Per and move into his head by reaching Miram (η Per). His bottom half descends from Mirphak to reach his waist at the double star δ Per. His legs then stretch lower through ε Per and Menkib (ξ Per), until you reach his feet at ζ Per and o Per. Finally, returning to Medusa's head, you'll find Gorgonea Tertia (ρ Per) which, like its neighbour, is a variable star.

Perseus is also home to the famous Double Cluster, two open clusters called NGC 869 and NGC 884, residing so close together they can be seen in the same field of view through binoculars and small telescopes. You'll find them beyond Miram at the top of Perseus' head. The cluster also sports the designation Caldwell 14. Amateur astronomers are likely to encounter the Caldwell catalogue on a regular basis. Caldwell is in fact the famous British astronomer Sir Patrick Moore, who created the list to complement the Messier catalogue after noting that the list missed out many of the sky's brighter objects. With 'M' already in use, he choose to use 'Caldwell', the first part of his double-barrelled surname.

To the right of the Double Cluster, close to the star φ Per, you will encounter the planetary nebula M76 (also known as the Little Dumbbell, it is similar in appearance to the Dumbbell Nebula in Vulpecula). Although it has a magnitude of 10.1, it has a reputation for being one of the hardest Messier objects to observe. Another equally elusive planetary nebula, the California Nebula, is to be found in Perseus' leg, very close to the star Menkib. Its low surface brightness makes it a real challenge.

Should these nebulae indeed prove too difficult to find, you could console yourself with another open cluster, M34. You'll find it close to the line between Algol and Almaak in Andromeda. At magnitude 5.5, you'll need binoculars to see it unless you are under dark skies.

There are several galaxies to hunt out here too. The first, NGC 1023, is a barred spiral galaxy not far from M34 and close to the border with Andromeda. At 38 million light years from Earth, we see it shining with magnitude 10.35. The lenticular galaxy NGC 1260 also lies about three quarters of the way along a line between v Per and Algol.

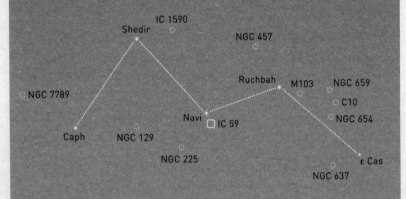

CASSIOPEIA

In autumn (October) looking north-east from Kielder Observatory

IC 1590

Shedir

NGC 457

NGC 7789

Ruchbah   M103   NGC 659

C10

Navi   IC 59   NGC 654

Caph   NGC 129

ε Cas

NGC 225

NGC 637

Stars   Mag 0 ✺   Mag 1 ✳   Mag 2 ✴   Mag 3 ✦   Mag 4 ·   Mag 5 ·   Clusters ⋰   Nebulas ☐

According to mythology, Cassiopeia was an African queen who boasted of her outstanding beauty.

During the summer months Cassiopeia tracks close to the horizon and is in an unfavourable position for easy observation. By September, however, she is on the climb in the north-east. With five of her brightest stars arranged in a distinctive 'W' shape, it is one of the easiest constellations to spot across the entire sky. Located right in the centre of the W is the otherwise nameless star γ Cas. The letter then branches out to the left with Ruchbah (δ Cas) and ε Cas, and to the right with Shedir (α Cas) and Caph (β Cas).

Around 5 degrees from Caph sits the star ρ Cas. Whilst it doesn't appear particularly bright in our sky (magnitude 4.5), it is actually one of the brightest stars in the Milky Way. It shines with the brightness of half a million suns and is so big that were it to replace the Sun in our solar system it would stretch out to twice the distance of Earth's orbit.

The constellation's familiar shape also aids in finding other well-known objects beyond the constellation's borders. First, use the three stars Ruchbah, γ Cas and Shedir to form an arrow. Following this arrow will take you very close to Polaris in Ursa Major, making Cassiopeia an excellent way to check you have indeed found the North Star. Similarly, follow another arrow formed by γ Cas, Shedir and Caph. This time you're taken in the opposite direction towards the neighbouring constellation of Andromeda and close to the famous Andromeda galaxy.

Thanks to its position on the band of the Milky Way, Cassiopeia is itself rich in deep-sky objects. The first is the open cluster M52, which is located along a line that extends from Shedir through Caph for just over the same distance again. At magnitude 5.0 it is easier to make out than the other Messier open cluster in Cassiopeia: M103. This shines at magnitude 7.4, so you'll need binoculars to find it near the star Ruchbah. It is one of the most distant open clusters known.

There are two other open clusters on offer: NGC 457 and NGC 663. Some observers have said that the former looks like the movie alien E.T. Also known as the Dragonfly cluster, one of its stars – φ Cas – is visible to the unaided eye. NGC 663 can be found just below the midpoint of the line between Ruchbah and ε Cas.

Two supernova remnants also reside in this constellation. The first one doesn't actually have a name, but is associated with one of the most famous stellar denotations – the explosion of Tycho's star. One of just a handful of supernovae in recorded history bright enough to be seen with the unaided eye, it appeared in the sky in November 1572. It is now named after the Danish astronomer Tycho Brahe, who wrote a beautifully named paper on the event, entitled *Concerning the Star, new and never before seen in the life or memory of anyone.* Brahe was a particularly eccentric character who lived in a castle on an island that now lies between Denmark and Sweden. As well as his astronomy, he is famous for losing part of his nose during a duel over mathematics, aged just twenty. He later mentored Johannes Kepler, although some historians accuse the latter of stealing credit for a lot of Brahe's work. The brightness of the supernova named after him is thought to have peaked at magnitude -4.0, making it brighter than Jupiter ever gets. Unfortunately, most of its light is no longer in the visible part of the spectrum and so cannot easily be seen with amateur instruments.

The other supernova remnant – Cassiopeia A – is just about visible in telescopes with 10-inch apertures and above. It is also the loudest source of radio emission in the sky. It is thought to have exploded around 300 years ago, but there are no records of a bright new star appearing at the time.

Finally, Cassiopeia offers up two elliptical galaxies. The magnitude-9.2 NGC 185 is located close to the border with Andromeda near the star o Cas. The second is the magnitude-9.3 NGC 147, which is only a few degrees away.

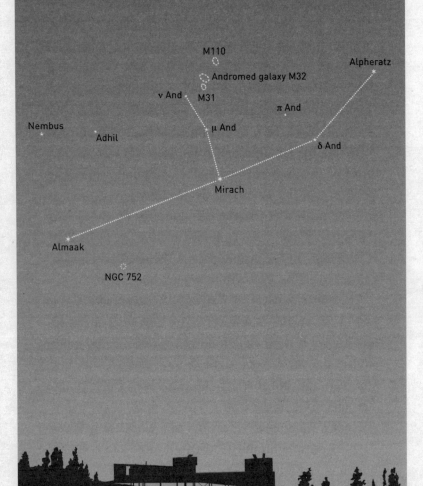

ANDROMEDA

In autumn (September) looking east from Kielder Observatory

M110

Andromed galaxy M32

ν And   M31

π And

Nembus

Adhil   μ And   Alpheratz

δ And

Mirach

Almaak

NGC 752

Stars   Mag 0 ✸   Mag 1 ✳   Mag 2 ✶   Mag 3 ★   Mag 4 •   Mag 5 ·   Clusters ⁖   Nebulas ☐

Some versions of the myth say that Cassiopeia conceded that Andromeda was the most beautiful creature on Earth – rather than herself. When it comes to the constellations, they're both a vision to behold. In the sky, the main line of Andromeda's body is formed by her brightest four stars. Her head is adjoined to the constellation of Pegasus (see November–December sky guide) at the star Alpheratz (α And). Move down to δ And to find her shoulder. By the time you encounter Mirach (β And) you've arrived at her waist. Continuing on to the Almaak (γ And) will bring you to her lower legs. Almaak is a beautiful double star that is a popular target for viewing. The primary is a 2.3-magnitude yellow star, with the secondary a dimmer magnitude-5.0 blue. Separated by a fairly sizeable 9.7 arcseconds, they are easy to distinguish. In fact, the blue star is itself a double – its secondary has a magnitude of 6.3.

Return to Mirach and head across her waist and you'll find μ And. Continuing on this line will take you to ν And, which represents the shackles used to tie her to the rocks.

The region above and beyond Andromeda's outstretched right arm is home to an asterism known as Frederick's Glory. Formed by the stars ι And, κ And, λ And, o And and ψ And, it was once part of the now defunct constellation of Frederici Honores, which celebrated the Prussian king Frederick the Great. That the constellations have been redrawn many times is further signified by the fact that Alpheratz used to be part of the adjoining constellation of Pegasus, where it was once designated δ Peg.

Andromeda contains one of the most famous of all the deep-sky objects – the Andromeda galaxy (M31). Despite being the nearest major galaxy to our own Milky Way, it still sits a staggering 2.5 million light years away. It contains more stars than our own galaxy and is twice as wide, but its enormous distance from us means that even to the unaided eye it appears as a fuzzy smudge among the stars. Find it by beginning with Mirach and then skipping up to μ And. M31 sits just beyond the next star, ν And. At 3 degrees across, it is six times wider than the Full Moon, so you won't get the whole galaxy in the field of view of a telescope. Binoculars are actually the best way to get an enhanced view. That said, you'll struggle to make out more than its bright central core.

It is worth pausing for a moment to think about the journey the light from the Andromeda galaxy has taken. The photons arriving on Earth today set off from the galaxy 2.5 million years ago. They have spent all those years trekking across space to get here. When they first departed, humans weren't even here on Earth. Instead, our ancestors – a species known as Australopithecus – were just starting to fashion stones into tools to kick-start the Stone Age. All of human civilisation has played out whilst the light entering your eyes has been journeying here. That fact alone makes it one of the most spectacular sights in the heavens.

Even better, it is only going to get more vivid as Andromeda and the Milky Way are on a collision course. Granted it will take at least another 3 billion years for them to merge, but as our neighbour gets closer and closer, it will grow in apparent size in the sky.

Elsewhere in the constellation you will find the open cluster NGC 752 sitting close to the line between Almaak and Mothallah (α Tri) in the adjacent constellation of Triangulum. It was discovered by Caroline Herschel in 1786. If M31 has whetted your appetite for distant galaxies, then it might also be worth leaving Andromeda and continuing into Triangulum to hunt down the Triangulum galaxy. It should be possible to make it out with the unaided eye as long as you're observing from the darkest of sites.

Andromeda's last remaining deep-sky object of note is the planetary nebula NGC 7662, nicknamed the Snowball Nebula. Amateur telescopes with apertures above 6 inches should be able to pick out the bright central star surrounded by a bluish disc.

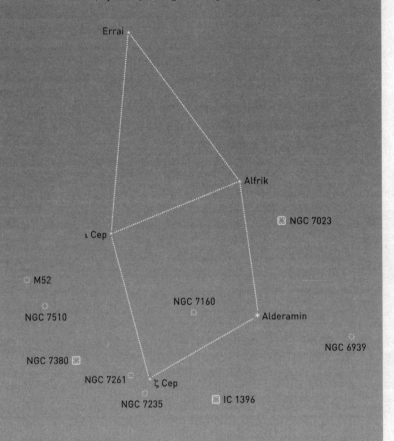

CEPHEUS

In autumn (September) looking overhead from Kielder Observatory

Errai

Alfrik

NGC 7023

ι Cep

M52

NGC 7510

NGC 7160

Alderamin

NGC 6939

NGC 7380

NGC 7261

ζ Cep

NGC 7235

IC 1396

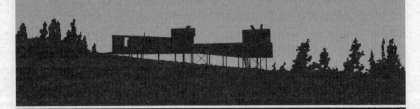

Stars Mag 0 ✳ Mag 1 ✱ Mag 2 ★ Mag 3 ⋆ Mag 4 · Mag 5 · Clusters ✵ Nebulas ☐

Cepheus was the king of Aethiopia in Greek mythology. Sitting above both Cassiopeia and Andromeda, the constellation Cepheus takes the form of a triangle of stars attached to a square – a sort of inverted house shape. To form one side of the square, start with his brightest star, Alderamin, before descending to Alfirk (β Cep). The triangle then peaks at Errai (γ Cep) before rejoining the square at ι Cep. Complete your journey by climbing towards ζ Cep and then returning to Alderamin.

The star δ Cep is particularly famous because it was the first Cepheid variable star discovered – hence the name of this class of star is used to help astronomers measure distances in space. It lies close to ζ Cep and those two stars form a small triangle with ε Cep. Like γ And, δ Cep is also a double star with striking yellow and blue components. Just above the line between ζ Cep and Alderamin you can find yet another variable star, this time known as the Garnet Star (μ Cep). It was named by William Herschel, who noted its rich, deep red colour. Its brightness climbs from magnitude 5.1 to magnitude 3.4 over a period of about two years. Not far from ι Cep sits the double star o Cep, although its colours are less striking than some other doubles.

Whilst Cepheus is devoid of Messier objects, there are still some deep-sky treasures to search for. NGC 188 is the closest open cluster to the North Celestial Pole (NCP), meaning it lies far from the plane of the Milky Way. This explains its particularly ancient age of 5 billion years. Usually open clusters are eventually torn apart by the gravitational pull of our galaxy. However, the lofty position of NGC 188 means that it has been able to persist. You'll find it within 5 degrees of the NCP, close to the line between Errai and Polaris in Ursa Minor.

Next up, search for the magnitude-9.6 Fireworks Galaxy (NGC 6946). It is in the unusual position of sitting exactly on the borderline between two constellations: Cepheus and Cygnus. A face-on spiral galaxy, nine supernovae have been observed to explode there in the last 100 years – far above the average rate for a galaxy. Find it by starting with Alderamin and heading for the star η Cep. Close by are the stars HIP 102216 and HIP 102011. Extending the line between them away from the main body of Cepheus leads you straight to NGC 6946.

Cepheus is also home to the Elephant's Trunk Nebula (IC 1396A), part of a much larger region of gas called IC 1396 sitting just to the side of the Garnet Star. The tower of gas is believed to be an area of current star formation. Whilst it can be hard to make out much detail through the eyepiece of a telescope, it is a popular target among astrophotographers where long-exposure images can draw out more detail.

Finally, although not visible to amateur astronomers, it is worth noting that Cepheus is home to a particularly powerful quasar – one of the universe's most violent outbursts, caused when matter spirals into a distant super-massive black hole. It is one of the most luminous quasars found to date, and the central black hole is thought to be 40 billion times more massive than the Sun.

Jupiter

Along with Venus, Jupiter is one of the easiest planets to spot in the night sky. That's mostly down to its enormous size: you could fit around 1,300 Earths inside Jupiter, and you could even fit all the other planets in the solar system inside it too. This huge surface area means it reflects a significant amount of sunlight back towards us, which explains its brightness despite being more than five times further away from the Sun than the Earth. Interestingly, Jupiter is only 318 times more massive than our Earth: this is because the planet is a gas giant, a planet made from hydrogen gas as well as helium.

At its brightest, Jupiter can get close to rivalling Venus

for brilliance in the sky. For this reason, the two planets often get mistaken for one another. Whilst Venus at its dimmest is always brighter than Jupiter at its most luminous, another way to distinguish them is to note how high the object gets in the sky. As Venus is the solar system's second planet, it never gets more than 47 degrees from the Sun. So if you think you're looking at Venus, but the object is higher than that in the sky, then it is probably Jupiter instead.

The planet's great size also means you don't need a huge amount of magnification to start seeing it in greater detail. Even a small pair of binoculars will reveal it as a round, planet-shaped disc rather than a point-like star. Slightly higher magnification, perhaps through a small telescope, will begin to show some structure to that disc. The first thing you're likely to make out is the planet's distinctive orange stripes, known as equatorial belts. There is a chance you might also be able to spot Jupiter's famous Great Red Spot. However, it could be on the other side of the planet at the time. Fortunately, you won't have to wait too long for it to return as Jupiter's rotation period of just under ten hours will soon carry it back round – Jupiter spins faster than any other planet in the solar system.

Binoculars or a small telescope will also reveal up to four bright, star-like objects in line with Jupiter's equator. These are the so-called Galilean moons, the biggest of Jupiter's hoard of at least sixty-three natural satellites. Named Io, Europa, Ganymede and Callisto, they were first seen by Italian astronomer Galileo Galilei in 1610 (see Chapter 2). In seeing objects orbiting around a planet other than the Earth, his observations effectively killed off the idea that our planet was at the centre of everything.

The joy of the Galilean moons is that they orbit Jupiter relatively quickly. Io, for example, takes approximately 1.5 days to complete one circuit. So even over the course of a week the configuration of the Galilean moons can change markedly. Moons will come and go from view as their orbit either takes them behind or in front of the planet. On occasion these moons will cast their shadows onto the main disc of Jupiter too. These events – known as shadow transits – can be seen by amateurs through large-aperture telescopes.

6

Infinite

'If you had to build a visitor attraction today from scratch, what could be better than the universe?'

OK, I admit it, this is a useful sound bite I sometimes feed to benefactors or journalists when we need to fundraise or promote the observatory. But it also has the advantage of being true.

It is 19 March 2015, the night before the last solar eclipse that will grace the skies of northern Europe and the Arctic for over a decade. I am at the observatory late at night with my team after finishing up another event. We are busily discussing tomorrow's grand showing. Every year there are at least two solar eclipses on Earth, although in some years there can be up to five.*

---

* According to NASA calculations, there have only been twenty-five years in the past 5,000 years that have had five solar eclipses. The last time this happened was in 1935, and the next time will be in 2206, when two solar eclipses will occur in December.

The last impressive solar eclipse that was witnessed in the UK occurred in 1999, when many cities were smothered in cloud and people struggled to see it. But solar eclipses are very special events. In astronomical terms, they are flukes, an observed phenomenon due entirely to a coincidence in the geometry of our solar system. How does the Moon perfectly cover the Sun during an eclipse? Well, only because our Sun is approximately 400 times bigger than the Moon, and the Sun is 400 times further away from the Moon. From our viewpoint on Earth, this coincidence manifests itself in the two objects lining up and appearing to be the exact same size, and so we get an eclipse.

Tonight the sky at Kielder has been pleasingly clear and the Milky Way has been ever-present. Dan Monk is one of our youngest members of staff. Unfortunately for him he used to be my next-door neighbour, which is how we met, but he has been looking at the stars since he was ten years old in his backyard. Night after night he would look up into the night sky, and now, fourteen years later, he knows the heavens from Kielder like the back of his hand; he can recognise most stars, constellations and objects with stunning accuracy. We had fielded lots of questions about the eclipse tonight and, as ever, Dan was more than adept at answering each one.

I watched proudly as he explained to one gentleman how our Moon is slowly moving away from us at a current rate of 15mm per year due to the effect of tidal interactions with the Earth. Dan gestured animatedly to describe how these tidal forces produce friction, which in turn slowly reduces the spin of the Earth. The conservation of angular momentum results in the Moon speeding up and thus moving away from us. He finished his answer with an awe-inspiring thought: billions of years ago the Moon would have appeared much bigger in the sky; billions of years in the future it will appear much smaller. He added that the next time the visitor looks at the Moon he should remember that we probably owe our very chance of existence to it: if the Moon did not exist, then our Earth would spin much faster; in fact, a day on Earth would probably be around seven hours long and existing on a planet that spins as fast as that would be very difficult for us and many other mammals.

The rest of the staff are still here too. A recent graduate in astrophysics, Hayden Goodfellow is every bit the nerd, but proudly so, feeding off every scrap of maths and scientific detail. Get it wrong and Hayden will tell you so; he also has an uncanny knack of breaking nearly everything in sight. Sitting opposite Hayden by the fire is Dr Fred Stevenson, a hippy cosmologist in his sixties, formerly from Durham University with a Ph.D. in cosmology. Fred is so laid-back; he thinks it's all just 'cool'. I love Fred; well, everyone loves Fred. Then there's Dr Sam James, a quantum physicist with crazy amounts of energy, who can and will fix everything and who just can't stay away from us, but we wouldn't have it any other way. It's a great mix to have Hayden break everything and then Sam repairs it, perfect. Finally there is Neill Sanders, one of our brilliant and tireless regular volunteers and tech gurus, who is always on hand to ensure the Kielder ship sails the celestial waves smoothly.

Together we go over some of the questions about the eclipse that we can expect to answer tomorrow; primarily, the different types of eclipses. Sometimes the Moon's disc only appears to graze the limb of the Sun, providing a partial solar eclipse. At other times the Moon can sit in the middle of the Sun, leaving a glowing ring around the outside (an annular eclipse). But occasionally it can block out the Sun entirely, leading to a spectacular total eclipse of the Sun. When the Moon has ghosted almost completely in front of the Sun, sunlight will only shine through the valleys on the edge of the Moon – an effect known as Baily's beads. The Diamond Ring effect follows, the last flash of light before a total eclipse. This totality can then last for a maximum of seven minutes and forty seconds. At this time we are treated to a very rare glimpse of the Sun's corona – its tenuous outer atmosphere that stretches out through our solar system. You might also be able to see solar prominences billowing out from the edge of the Sun. Stars, too, may be visible in the area around the darkened

Sun. During an eclipse in 1919, English astronomer Arthur Eddington famously used the positions of these stars to prove Einstein's General Theory of Relativity.

Tomorrow's event will be a total eclipse in very northern places, such as off the coast of the Faroe Islands. The duration of totality there will be nearly three minutes. From the UK it will be a partial eclipse, and visible for a lesser time, but it will still be impressive – we are expecting just over 80 per cent obscuration, which will mean the sky will get noticeably darker and the temperature will temporarily drop in the absence of the Sun. In wild places like Kielder, the local animals may fall silent as they are tricked into thinking it is dusk.

Over the last few months at Kielder, with my staff I have formulated a plan to show as many people as possible images of this eclipse in towns and cities up and down the country; we want to bring this rare spectacle to life. Of course, you should never look directly at the Sun – even during a total eclipse – as you run the risk of sunlight re-emerging. The safest and simplest way to get around this is to build a pinhole camera – a piece of card with a hole in it – to project the image of the Sun onto another piece of card. Alternatively you can buy special eclipse glasses which filter out the harmful radiation or, in my case, a pair of solar glasses, with in-built protective film. But because this coming eclipse will not be visible to everyone in the UK, we have come up with another solution. We have hired giant screens and distributed them across the UK. The idea is simple: from the observatory we will transmit live images of the eclipse to the screens; we have even set up a huge screen in the middle of Newcastle city centre. The aim is for locals to gather there with office workers and shoppers. A row of deckchairs will be laid out for people to lie on and wait for the spectacle: a combination of seaside resort and cosmic cinema. Kielder Observatory volunteers, along with students from Durham and Newcastle universities, will be on hand to offer advice on the physics of the eclipse. All in all, we predict that our video footage will be beamed to over 1 million people. It will be available to watch on the web, and shown on billboard screens and road-side hoardings up and down the UK by motorways and at service stations. We've spent thousands of pounds ensuring we have the correct means to show as many people the eclipse as possible. It is all set; we just need clear skies.

It is now past midnight, but before we pull out the camp beds – no one is leaving here tonight – we check the weather patterns for the final time and discuss our plans. Plan A assumes clear skies at the observatory so that we can beam the images around the world. Easy. If it is cloudy, Plan B is a bit more complicated. It will mostly consist of Neill getting into his car and driving somewhere else where the skies are clear, filming the eclipse and transmitting the video files as a stream from there. Neill has devised a clever way in which he can reduce the size of the images to a tiny file size, so that we can upload it on the hoof. With a fast refresh rate, it should appear across the UK as real-time video – genius! We hope. In the morning we will have television and radio crews from the BBC and ITV standing by at the observatory, as well as a full house of guests ready to witness this potentially once-in-a-lifetime event. I know I'll sleep light tonight, if at all.

I wake up regularly every hour until 5 a.m. and check the forecasts immediately on my laptop. The skies are going to be cloudy. It has to be Plan B. We need to get Neill mobile. All of the staff wake up, as if programmed together once one person raises their head out of the sack. An hour later, still groggy, we are all busy assembling the kit. Under Neill's supervision, it is checked and double-checked; fortunately everything works. We pack the telescope and cameras carefully into Neill's car and begin to plan his route. After scouring the local weather forecasts, our best guess for Neill seems to be to head east. At this point as if on cue a crew from the BBC arrives at the observatory and proposes a solution: Neill could drive to the BBC offices in Newcastle and set up his kit on the roof. He'll have an unimpeded view and he can also use the building's Wi-Fi. Bingo – we go for it. I wave Neill off.

One hour and thirty minutes passes. Nothing from Neill. If this is messed up, it will be a very public embarrassment; thousands of people will be disappointed. My chest tightens. I try to call his mobile, but he doesn't pick up; he must be out of range. Where could he be? But I don't have much time to worry as, despite the clouds, we are still setting up the observatory for the eclipse, just in case. Dan and Matt scuttle in front of me, readying our scopes and cameras; volunteers prepare the presentation in the classroom; I'm project-managing the team and screens across the country. My phone rings, but it isn't Neill. 'Gary, there is no image.' It is the team in Newcastle city centre. Hundreds of people have gathered there, the Kielder Observatory logo plastered across the hoardings by the giant screen, but no image of the sky – just noise and scrambled signal. My chest tightens again. Where is Neill?

'Find a NASA stream, quick.' As a last resort, showing someone else's stream will be better than complete failure. I give Dan the brief and he is back at the desk within seconds with several options. I wonder if Neill has been in an accident; I push the thought out of my mind. Knowing him as well as I do – a man who is as driven as I am – I'm still hopeful he will find a way to pull this off somehow. He has to . . . I call BBC Newcastle for the umpteenth time, but he still hasn't arrived. I begin to resign myself to defeat. 'Out of interest, how's the weather there?' I ask, almost on autopilot. There's a pause on the line. 'You won't like it, Gary. It's cloudy as hell here.' I pause. I think they're expecting me to shout and swear. But it's actually the opposite. There might be a glimmer of hope. If I was Neill, and it was cloudy in Newcastle, I wouldn't go there either. Neill might just be travelling to clear skies somewhere else. However, there is still no call, and it is past 8.30 a.m. now. Something tells me Neill is off on a Plan C. The eclipse is visible from now until 10.40 a.m. and the best view will be at around 9.30 a.m. Time is running out.

*

An hour and a half before, Neill had packed his Audi A4 with the portable HEQ5 telescope mount, the Coronado solar telescope, the Watec 120N light-sensitive camera and an assortment of batteries and cables. During the eclipse, this sophisticated mobile set-up should reveal in vivid detail the Sun's surface, including sun spots, filaments and prominences.

Having made his way down the observatory track, dodging TV crews and radio vans coming up the other way, Neill reached the main road and made a key decision. He had a hunch Newcastle would still be cloudy in an hour's time, so he took the forest road east that links Kielder to the A68, north of Otterburn.

A track more suitable for four-wheel-drive vehicles than his sporty saloon, the road was undulating, full of potholes and definitely not designed for his low-profile tyres. It took fifty minutes to drive twelve miles. There was no phone signal either and, apart from two timber lorries he was briefly stuck behind, he saw no one else in all of that time. If he got a puncture, he would be in trouble, but there were brightening skies visible to the east which kept his spirits up.

Just before 8 a.m. Neill finally drove back into phone signal coverage and immediately called a friend, a fellow astronomer who was at home nearby in Rothbury. 'It's looking relatively promising. You're welcome to come here and set up – we're having an eclipse breakfast party.' With the promise of better skies and a bacon butty, Neill continued to head east towards the coast, even though time was ticking on.

It was about fifteen minutes later when Neil arrived in Rothbury. Thankfully there were now plenty of breaks in the cloud and the morning was bright. With around a thousand people gathered in Newcastle, Neill was feeling the pressure. He drove straight to the village green to set up the telescope and camera, but there was no 3G internet connection to upload the video for the live stream. For the next ten minutes Neil drove around Rothbury desperately, holding his mobile phone out of the car window looking for a connection so that he could create a local Wi-Fi network for his laptop. Nothing. Rothbury was like the Bermuda triangle. He headed out of the village into the hills for one last shot. After another few minutes he felt his phone vibrate. It was Dan at the observatory.

*

Back at Kielder, the eclipse has started but we can't see it. I have that sinking feeling. Months of preparation wasted – all for naught. I hear a clamouring outside. Great, what else could go wrong now?

'Get Gary!' I hear Dan's shout first, then see him running towards me. He nearly throws the phone at me.

'Hiya mate.' It's Neill.

'Where the hell have you been? Where are you?'

'I drove north, to escape the clouds. Nearly set up now and ready for a test. Can you stay near the phone, Gary?'

Completely unaware of our worries, Neill had ended up on a hilltop two miles outside of Rothbury, which turns out to be a fantastic location for today with a clear easterly horizon. The site has even temporarily become a hotspot for local eclipse-spotters; Neill is there with a handful of nearby residents, including a British Gas man, who is helping him to set up.

A few minutes later, with one arm raised in the air to maintain the best 3G signal, Neill positions the telescope by the roadside, connects the camera to the laptop and beams out the resulting video image of the stunning eclipse around the world. He later said that it was a surreal feeling connecting so many people to such a wonderful experience as the solar eclipse, from the side of the road.

Inside the classroom at Kielder, we are elated. As are the public in Newcastle. I'm on the phone with Patti, our office manager who is running the show, as I hear an almighty roar, followed by clapping and shouts ringing down the line: Neill's video stream is up and running. Moments earlier Patti had been frantically trying to allay their fears; now they are jubilant. I look at our monitor and upload the image. Sure enough, there it is: the Moon passing in front of our home star. The sky is darkening as the last light of the Sun is briefly obscured.

I rush outside the observatory to tell the visitors to come in and see these special pictures. I needn't have bothered. Why? Sod's law. The cloudy forecast for Kielder has disappeared. It is crystal-blue skies and the guests can see the eclipse just fine now with their own eyes as the heavens above us darken and the temperature starts to drop. Cameras are running and telescopes with solar filters are tracking. The staff are explaining the physics in hushed tones. Most people are smiling in their special eclipse protective glasses; I share my own pair around. Standing together in the strange twilight, witnessing this once-in-a-decade event, I feel so humbled and fortunate. Millions upon millions of conditions had to be just right for today to happen. But that is exactly the nature of astronomy – always be prepared for the unexpected.

*

In the year since that event, Kielder has continued to go from strength to strength. We now employ nine full-time members of staff, brilliantly aided by twenty-five volunteers and eight dedicated trustees. We are an eclectic group, to put it mildly, composed of both amateur and professional astronomers, quantum theorists and astrophysicists, an ex-brickie, an ex-car salesman and former call-centre operators. But like a sci-fi alliance of vigilantes, somehow we have come together and pulled it all off. We now run stargazing events nearly 365 days a year: our aim is to enthuse, educate and inspire every visitor.

Of course, we could not have done any of this without the unwavering support and endless curiosity of our visitors, as well as the advances in science and technology that have captured the public's imagination and made the world interested in space again. In an age of technology, we are hurtling towards an increasingly advanced perspective of the universe. Since the observatory opened in 2008, in no particular order, scientists around the world have discovered the Higgs boson, the particle at the end of the universe responsible for giving mass its mass; we have sent probes to the distant dwarf planet Pluto, which have beamed back to us in unprecedented resolution images that vividly depict the tiny world's surface for the first time; we have discovered gravitational waves, resulting from the merger of two massive black holes over a billion light years away, the effect of which has set the fabric of space wobbling in its wake. Regularly we send men and women into orbit on the International Space Station to conduct experiments; we build bigger and more sensitive telescopes capable of peering into the dim and distant history of the universe. Slowly, as we attempt to piece all of this together, we take further steps on our noble quest to understand more about the true nature of reality.

Today it's even 'cool' to be an astronomer, a technologically advanced explorer of remote and otherworldly shores. However, the astronomer's reliance on technology undoubtedly presents risks. When I was sitting in front of one of the powerful robotic instruments at the Chilean observatory, I thought back to my days as a child using my brother's small telescope. Our eyes used to do the important observational work; now it is computers downloading mounds upon mounds of data, often to be sifted through by even more powerful computers stored inside faraway research institutes. Science is by its very nature empirical, but is the romance that sparked our imaginations in the first place being lost? In Chile I found myself in control rooms full of banks of machines, the quiet burr in the background of fans cooling hard drives. Though inflatable planets still hung from the ceiling, a timely reminder of the youthful playfulness of the scientists hard at work, their eyes were glued to their screens. They didn't look as if they had seen the real sky for days. Is their inspiration, their ability to dream about our universe, getting lost in the technical jungle of scientific progression? And as machines get more sophisticated, will the need for scientists disappear altogether – to be replaced by AI with mind-bending computing power? Unlike during the days of William Herschel and Edwin Hubble, when human observation was everything, is it now all about the electronic data, the language of 0 and 1?

I share some of these concerns, which is why I always encourage beginners in astronomy to learn to navigate the sky with their eyes first, and to master a basic telescope before using one that can automatically find a celestial location. I hope that this book may be of use to some readers who do wish to find the sky for themselves. But I know, too, how technology can make the universe even more beautiful. Someone, a guest at Kielder Observatory, once asked me if the more I learned about science, the less spiritual I became as a person. My answer was an emphatic 'no'. Astronomy has made me more spiritual than I ever imagined I could be. To live in a potentially infinite universe, where the properties of four funda-mental forces – gravitational, electromagnetic, strong nuclear and weak nuclear energy – govern the mechanics of everything, where I can look back through the history of time by simply looking into the night sky, well, that for me is as profound as it gets.

Telescopes are like books in many ways. They are founts of discovery and wonder; they open up the depths of the universe to interpretation and inquiry. Turn the page to the Moon to see the craters and valleys; now turn it to the giant planet Jupiter and dream of Galileo observing it in 1610; flip to the faraway galaxies strewn across the cosmic web, so distant they test the limits of our understanding and force us to contemplate the evolution of our universe. We just have to be willing to listen to the universe – like a story – and to let our human spirits soar.

Take M51, also called the Whirlpool. M51 is a large spiral galaxy that consists of gigantic clouds of collapsing dust and gas condensing into new stars. Located over 20 million light years away from Earth, the combined light of its 300–400 billion suns merges with a nearby dwarf elliptical galaxy. Observing M51 is like looking at a penny-farthing bicycle: a big wheel and a little one next to it, representing the cores of the two galaxies. Bright streamers of stars are visible as gravity pulls them away from the merger. But to really admire M51's beauty we need to use the world's most advanced telescope to help us. The Hubble Space Telescope was launched into low orbit in 1990 aboard the space shuttle *Discovery*, where it has remained ever since, circling some 560 kilometres above Earth, looking deep into the universe. The images of M51 taken by Hubble's six cameras and sensors via its 8-foot-diameter (2.4 metres) mirror, utilise ultraviolet and infrared light, and are a visual delight. When we show visitors reproductions of the images in the classroom at Kielder, people often think the images have been enhanced. They are awestruck by the spiral galaxy's vital pink colour, the tell-tale regions where infant stars are being forged. They are impressed, too, by the blue light from the hot massive stars forming around the spiral arms. Mixed in between these areas is the so-called 'dust': the polycyclic aromatic hydrocarbons (PAHs) which are the building blocks for future generations of stars. To observe M51 is to observe creation and regeneration – life itself. Without the technology of Hubble, and the rest of the world's best ground-based telescopes, we would never be able to see such richness and colour. It is the touchpaper for people's imaginations.

I sometimes ask guests at Kielder to try to imagine what the view of M51 would be like in the night sky of other planets orbiting stars inside of this grand galaxy. Try to picture, just for a moment, what the vast bright clouds of gas strewn across these alien skies would look like. Fields of stars would merge spectacularly with the galaxies overhead; a sight we can only dream of. Sometimes, well quite often actually, guests come to visit us in couples. Invariably, one of the partners likes astronomy more than the other, who has come along as a romantic gesture or as a mark of solidarity. After half an hour, the latter partner normally starts to display the usual behaviour: restlessness, fidgeting and feeling the cold. But M51 invariably comes to the rescue – it's an astropessimist's antidote. One look and it's 'Oh my God! What is that?' They have been converted.

Seeing the Kielder Observatory develop and grow over the last decade has been the single greatest professional achievement of my life. Galaxies like M51 in many ways remind me of the journey I have been on: how through the laws of physics wonderful things can happen, but that so many factors need to be synchronised, including a great slice of luck, for it all to work. I hope Kielder will always be a visitor attraction, a safe place to gaze up into the night-time sky; where kids can get wide-eyed and excited about the universe and where eighty-year-old ladies can see Jupiter for the first time.

My ultimate wish for the observatory is to take it one step further: that visitors may one day be so inspired by the thought of these cornucopias of stars that they will go off and make new discoveries of their own. It is always heartening to receive letters from parents whose children, after visiting us, have taken more of an interest in science or astronomy. I have to mention Jasmine Evans, who first came to us as a visitor aged fifteen, and then returned when she was sixteen as a volunteer. Jas lived for the observatory; within her tiny figure lay a passion for astronomy and a well of knowledge that radiated outwards like the rings of Saturn. Guests would often comment on her enthusiasm and dedication to the facility. Jas is now at university studying physics; one day it's her dream to be an astronaut and fly to the Moon or further afield. I wouldn't bet against her achieving those ambitions. My son James has also followed in the family business. Soon he joins the UK astrobiology group at Edinburgh University, working with NASA to look for signs of life in volcanic rocks: a potential precursor to a method for detecting life on Mars. This is truly a golden age for humanity to discover the boundless possibilities of our universe.

As I have said before, you don't need a Ph.D. in astrophysics to appreciate the night sky. However, we must support and appreciate the scientists with Ph.D.s as they are the ones who have committed their lives to furthering our understanding. In the coming years, we hope to provide more facilities at Kielder for youngsters looking to take their interests in stargazing to the next level. I want to build an astronomy village, a world-leading centre for excellence in astronomical outreach that will consist of an eighty-seater planetarium, a 1-metre aperture telescope, a solar observatory and an observing barn for school children. We are also hoping to set up partnerships with universities on a series of initiatives across the region, including building a new visitor centre that will provide local opportunities for business growth in science and renewable technologies. From the land to the sky, from the reservoir to the forests, the centre will be designed to investigate the local flora and fauna along with the stars; it will be a place where students can learn about our natural world. A second centre is planned to be built in the city of Durham, which will celebrate and promote STEAM

(Science, Technology, Engineering, Art and Mathematics). There, dedicated staff and laboratories will enthuse and educate students in these disciplines, using cutting-edge technology to inspire the next generation of scientists.

For me, and for now, I will continue to do what I have always done. I am a bricklayer by trade, an astronomer by choice. All I really want is for others to have a chance to enjoy what I have been lucky enough to experience. The motto for the observatory now is 'Infinite Inspiration'; maybe not as catchy as 'Who Stares Wins', but about right – and it looks good on our mugs. I've recently adopted a puppy, who we've named Lyra, after the constellation. She's tearing the place up and driving my partner Sarah and me mad. Naming our pet after my job might be a step too far – even the Sunderland Astronomical Society might take issue with that. But then again, I don't really consider astronomy a job. Taking my four adult children to the observatory, as well as friends and colleagues who only ever knew 'Gaz the brickie', fills me with pride that the universe, or fate, call it whatever you like, reached out and grabbed hold of my life. It pointed me towards Kielder and taught me to tell stories of faraway worlds and galaxies; now I live to dream.

I never did get to become a professional scientist, but as a tour guide to the universe I am as close as

I could ever have hoped to have been. In 2012 I was deeply honoured that Sir Arnold Wolfendale, a former Astronomer Royal, presented me with an honorary Masters in Science from Durham University at a ceremony at Durham Cathedral. I will not forget his kind words: 'Gary Fildes did not have a university education, indeed he left school at 17. Before we go 'tut-tut' we remember that the First Astronomer Royal, John Flamsteed, did not do either. Nevertheless, he made rather a good job of running the Royal Greenwich Observatory from 1675 to 1719, when Edmond Halley took over. Gary, as Founder Director and Lead Astronomer at Kielder, has worked inordinately hard to get the show on the road. Our hero, Flamsteed, would have been proud of him.'

I have no regrets, ever, but I wish my dad could see me. He never will, but hopefully this book may inspire you to look up and appreciate what you have here and now, if only for a second. We are all part of this awe-inspiring universe; just look up and let it in. You never know what may happen. Look up, please.

November–December
Sky Guide

*Pegasus – Pisces – Aries – Aquarius –
Saturn – International Space Station and
iridium flares*

PEGASUS

In autumn (November) looking south-east from Kielder Observatory

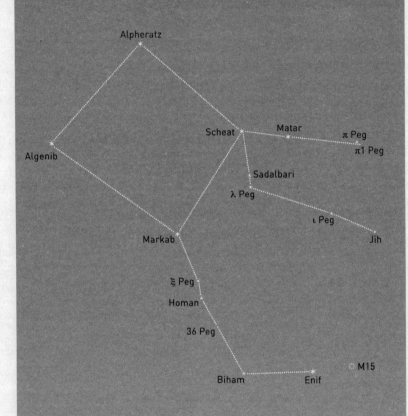

Alpheratz

Scheat • Matar • π Peg
π1 Peg

Algenib

Sadalbari
λ Peg

ι Peg

Jih

Markab

ξ Peg
Homan

36 Peg

Biham • Enif • M15

Stars   Mag 0 ✹   Mag 1 ✴   Mag 2 ✳   Mag 3 ★   Mag 4 •   Mag 5 ·   Clusters ☼   Nebulas ▢

The winged white horse is one of the most recognisable creatures from Graeco-Roman mythology, so it is no surprise to find Pegasus among the stars. At the end of November he is found due south at approximately 7 p.m. It is worth noting that the horse appears upside down to most northern hemisphere observers.

By far the most distinguishable part of this constellation is the Square of Pegasus – four stars arranged in a giant quadrilateral spanning many degrees of sky. Only three of its stars are part of Pegasus, however – the fourth is Alpheratz in Andromeda. Start here and head towards the horizon until you find Algenib (γ Peg), then run westwards across the ridge of the horse's back to reach Markab (designated α Peg despite not being the brightest). Climbing up to the red star Scheat (β Peg) will allow you to complete the square by returning to Alpheratz.

All that remains is the horse's head and legs. The neck joins the square at Markab and descends through ξ Peg and Homan (ζ Peg) to reach his eye at Biham (θ Peg). Finally his nose extends outwards from here to Pegasus' brightest star, Enif (ε Peg). To find the horse's two front legs, you need to start at Scheat. One leg is formed by drawing a straight line to Matar (η Peg) and on to π Peg. The second leg is bent, starting again at Scheat, but moving down to Sadalbari (μ Peg) and Sadalpheretz (λ Peg) before extending outwards in a straight line to ι Peg and the double star κ Peg. Sadalbari means 'luck star of the splendid one' in Arabic.

Between Markab and the stars Sadalbari and Sadalpheretz lies the magnitude-5.5 star 51 Pegasi. In 1995 it became the first Sun-like star to have a planet detected orbiting around it.

By far the stand-out deep-sky object in Pegasus is the globular cluster M15. With over 100,000 stars all packed into a group 175 light years in diameter, it is one of the densest globular clusters orbiting the Milky Way. At 12 billion years old, it is also one of the most ancient. Shining with magnitude 6.4, binoculars should be sufficient to get a glimpse of it. Telescopes with apertures above 6 inches will be able to begin resolving individual stars. This remarkable cluster is also home to many variable stars, as well as the first planetary nebula discovered inside a globular cluster. You'll find it just beyond the star Enif at the tip of the horse's nose.

Elsewhere in the constellation, start with Scheat and head along the horse's lower (straight) leg past Matar and about halfway to π Peg. This is the location of the orange, magnitude-6 star HIP 110873. From here, head down at a right angle to the horse's leg to reach 38 Peg.

If you continue on this line you will pass close to Stephan's Quintet. Named after nineteenth-century French astronomer Édouard Stephan, it is a compact group of five galaxies all interacting gravitationally with one another.

Within a degree of Stephan's Quintet is the magnitude-10.4 spiral galaxy NGC 7331 (Caldwell 30). Often referred to as the Milky Way's Twin, it sits 40 million light years away and was discovered by William Herschel in 1784.

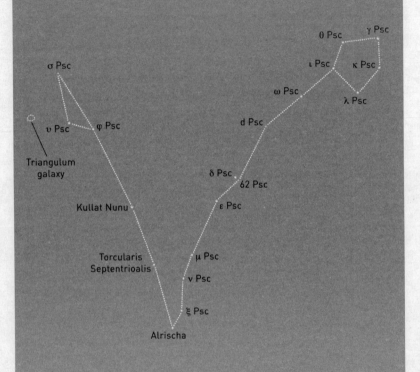

PISCES

In winter (December) looking south from Kielder Observatory

σ Psc

θ Psc   γ Psc

ι Psc   κ Psc

υ Psc   φ Psc

ω Psc   λ Psc

Triangulum
galaxy

d Psc

δ Psc

62 Psc

Kullat Nunu

ε Psc

Torcularis
Septentrioalis

μ Psc

ν Psc

ξ Psc

Alrischa

Stars Mag 0 ✹ Mag 1 ✷ Mag 2 ★ Mag 3 ⋆ Mag 4 · Mag 5 ·   Clusters ⬡   Nebulas ☐

One of the twelve signs of the zodiac, the constellation Pisces resembles two fish in the sky joined together by a long cord. Pisces is a relatively dim constellation, but is particularly noteworthy as home to the First Point of Aries, the place where the Sun crosses the celestial equator to mark the beginning of spring. Thanks to the precession of Earth's axis, the point has moved into Pisces since it was first identified.

The constellation starts with the head of one of the fish just under Mirach in Andromeda and then darts off towards Cetus before skirting back underneath the Square of Pegasus to emerge with the head of the second fish almost as far across as Markab (α Peg). Unsurprisingly, this long journey sees the main body of the constellation pass through more than a dozen bright stars.

Let's start at the head near Andromeda. It is formed by the stars σ Psc, υ Psc and ϕ Psc. The body of the first fish then drops down towards Kullat Nunu (η Psc) before moving into the cord between the fish at Torcular (o Psc) and Alrescha (α Psc). The cord then tacks up towards Pegasus, passing through ε Psc, υ Psc, μ Psc, Revati (ζ Psc), Kaht (ε Psc), Linteum (δ Psc), d Psc and Vernalis (ω Psc) to reach the base of the other head at ι Psc.

Moving clockwise around this second head takes you in turn through θ Psc, γ Psc, κ Psc and λ Psc. The only remaining star of note is Fum al Samakah (β Psc), which sits just beyond this head and is sometimes depicted as the fish's mouth.

Within this constellation's borders you'll spot the Messier galaxy M74. Find it by starting with Kullat Nunu and locating the stars 105 Psc and 101 Psc not far from it. M74 forms the apex of a triangle with the line between those two stars at its base. It sits 32 million light years away and has two distinct spiral arms. Like our Milky Way, it contains at least 100 billion stars. We are seeing it face-on, which you might think would make it easier to see. However, its low surface brightness has led some to consider it the most difficult Messier object to spot. Low magnification is a must if you are to stand a chance of picking it out.

You'll also find a pair of colliding galaxies – NGC 7714 and NGC 7715 – within the quintet of stars making up the head nearest to Aquarius. Start with λ Psc and journey diagonally across the head towards γ Psc and before long you'll encounter the 16 Psc. The galactic duo sit right next door. At a combined magnitude of 12.2 they are a faint target, but are also a stunning example of the structures that can be created by the gravitational interaction between galaxies.

The constellation is also home to one of the newest meteor showers on record: the Piscids. It is associated with the comet 46P/Wirtanen, and the Earth first crossed its debris stream in 2012 when southern hemisphere observers saw about a dozen meteors fly from Pisces.

ARIES

In autumn (November) looking south from Kielder Observatory

35 Ari

41 Ari

Hamal

Sheratan

Mesarthim

Stars   Mag 0 ✴   Mag 1 ✳   Mag 2 ✶   Mag 3 ✦   Mag 4 ·   Mag 5 ·   Clusters ⁙   Nebulas ☐

Aries, the Ram, was the animal from which the Golden Fleece was obtained. Another of the twelve signs of the zodiac, Aries can be found between the constellations of Taurus and Pisces (and below Andromeda and Triangulum). Its brightest star is the magnitude-2.0 Hamal (α Ari) which sits at the ram's neck. His eyes are often depicted as Sheratan (β Ari) and Mesarthim (γ Ari). Boteïn (δ Ari) is adjacent to the animal's hind leg, so far down it is almost encroaching on Taurus' territory. Nearby is the seemingly unremarkable star 53 Arietis. However, its unusual nature was revealed when astronomers measured its speed. Travelling at 48 kilometres per second, it is hurtling through space compared to its neighbours. It is believed to have been ejected from the Orion Nebula – two constellations away – some 5 million years ago and has been fleeing ever since. The stars AE Aurigae and μ Columbae appear to have had similar trajectories. Perhaps they were in binary systems and their companions went supernova.

Mesarthim is actually a double star with components of similar colour and brightness and a separation of just under 8 arcseconds. It became one of the first doubles to be resolved through a telescope when it was discovered by Sir Isaac Newton's adversary Robert Hooke in 1664.

There are several other doubles in Aries too, although none to rival some of the other more colourful doubles elsewhere in the sky. You can choose between the magnitude-5.15 ε Ari in the Ram's hind leg, λ Ari not far from Hamal, or π Ari at the hoof of the other hind leg to ε Ari.

Like Pisces, the deep-sky objects in Aries present a challenge, with no Messier objects on offer. The first thing to look out for is NGC 772, located not far from Sheratan and roughly between the stars HIP 9379 and HIP 9248. A 11.1-magnitude spiral galaxy, it is twice the size of the Milky Way, but unlike our galaxy it has no central bar.

Moving back up to Sheratan, you'll find the galaxy NGC 722 right beside it, along with the galaxy NGC 719 on the way towards Mesarthim. On the opposite side of Sheratan to NGC 772, close to the double star 1 Ari, sits a group of six faint galaxies all between magnitudes 12 and 14. They are: NGC 695, NGC 697, NGC 694, NGC 691, NGC 680 and NGC 678. The last two are a pair of galaxies only around 200,000 light years apart, one elliptical and one spiral.

AQUARIUS

In autumn (November) looking south-west from Kielder Observatory

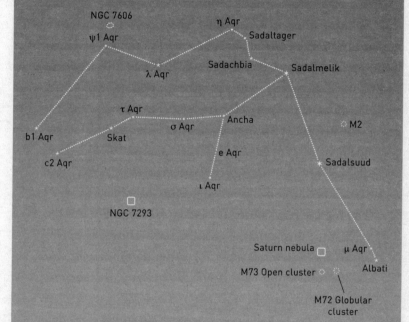

NGC 7606

ψ1 Aqr

η Aqr

Sadaltager

λ Aqr

Sadachbia

Sadalmelik

τ Aqr

σ Aqr

Ancha

M2

b1 Aqr

Skat

e Aqr

c2 Aqr

Sadalsuud

ι Aqr

NGC 7293

Saturn nebula

μ Aqr

M73 Open cluster

Albati

M72 Globular
cluster

Stars   Mag 0 ✸   Mag 1 ✶   Mag 2 ✦   Mag 3 ★   Mag 4 ·   Mag 5 ·   Clusters ⁛   Nebulas ☐

The constellation of Aquarius is often thought to represent a handsome young boy from Greek mythology who was transported to Mount Olympus to act as water-bearer to the gods. From Aries, if you skip across Pisces, you will end up at Aquarius. It is the tenth biggest by area of all eighty-eight star groups; consequently there are many stars here to grapple with.

Let's start with Sadalmelik (α Aqr) in the water-bearer's head. It is yet another example of the Bayer designation not matching the constellation's brightest star. From here his chest descends to Ancha (θ Aqr). The top of his right leg is marked by ι Aqr, but the rest of that leg is not marked out by stars. However, his left leg contains several. The first is the double star σ Aqr, before continuing on through Skat (δ Aqr) to reach his left foot at c Aqr.

Returning to Sadalmelik, we can fill in his right arm by heading towards Aquarius' true brightest star Sadalsuud (β Aqr) and ending up at the hand stars of μ Aqr and Albail (ε Aqr). Finally, we need to draw the stream of water with which the character of Aquarius is most associated. The cup he is carrying in his left hand is formed by the stars Sadachbia (γ Aqr), the double star ζ Aqr and then finally η Aqr. The water pours from here down towards his feet, zig-zagging through the stars Hudoor (λ Aqr), and the doubles ψ Aqr and b Aqr.

Hidden among this plethora of stars sit two globular clusters: M2 and M72. The first is located above the halfway point between the stars Sadalmelik and Sadalsuud, not far from the magnitude-6.2 star HIP 106758. One of the largest known globular clusters, it was discovered in 1746. M72 was first spotted thirty-four years later beneath Albali, close to the star HIP 102891. The open cluster M73 isn't far away either. It sits in a line of equally spaced out objects with the stars HIP 103728 and HIP 103640.

There are some cracking planetary nebulae here too. The Saturn Nebula (NGC 7009) gets its name from its distinctive side lobes, which resemble the famous rings of the solar system's second largest planet. Through a small telescope you might be able to make out its yellow-green colour. Find it from M73 by skipping up to HIP 103834 and then on to HIP 103801, before bending round towards ν Aqr. You'll see NGC 7009 before you get there.

The more famous Helix Nebula is located in the water-carrier's legs. Start with Skat in the left leg and head to his right leg through the stars g Aqr and υ Aqr. You're now in the right region to see one of the nearest planetary nebulae to Earth. In fact, it appears larger in the sky than any nebula of its kind. It is often nicknamed the Eye of God thanks to the way it appears in photographs taken through professional telescopes.

Galaxies within the constellation include NGC 7727 and NGC 7252; however, they are both quite faint and not as spectacular as some of the other galaxies found elsewhere in the sky.

Saturn

This planet is far and away the most popular target for newbie astronomers – everyone is keen to drink in the splendour of those beautiful rings. The good news is that Saturn is definitely visible to the unaided eye, making it easier to point a telescope at. However, as it is not as big as Jupiter and is more distant, it is smaller and fainter in the sky than its neighbour – it never stretches more than 22 arcseconds across. By contrast, Jupiter can reach more than double that. Saturn reaches a maximum magnitude of -0.5 when it is at opposition and the rings are arranged favourably to reflect the maximum possible light back to Earth. This last happened in 2003 and will happen again in 2018. At its dullest, the planet drops to magnitude 1.5.

You don't need much magnification to see the rings for yourself. Beware, though, that sometimes the orientation of Saturn means we are viewing the rings edge-on and so they become temporarily invisible, as they did back in 2009. Even through binoculars you'll notice that the planet does not appear round, but instead seems to have 'ears' poking out on either side. That's exactly what Galileo called them when he first saw them in the early seventeenth century. By the end of that century Christiaan Huygens had figured out that they were actually a ring system around the planet.

Those rings will begin to become distinct from the planet with magnifications above twenty-five times – i.e. with the smallest of telescopes. If the rings are orientated right, you might also be able to see Saturn casting its shadow onto them. You should also notice that the planet is significantly fatter at its equator than at its poles.

Boost your magnifying power to at least 100 times and you might start to notice there isn't one ring at all – there is at least one division between several rings. The most obvious of these gaps – roughly halfway out – is called the Cassini division. It is thought to exist because one of Saturn's moons – Mimas – is pulling on the ring material.

Speaking of moons, even the smallest of telescopes will reveal Saturn's largest: Titan. After Ganymede it is the second largest moon in the solar system and larger than the planet Mercury. Saturn actually has at least sixty-two moons, and you can see half a dozen or more with telescopes with apertures around 10 inches. The following moons all have magnitudes of 10 to 12 inches:

Mimas, Tethys, Rhea, Hyperion, Enceladus, Dione, Titan and Iapetus.

Those with large-aperture telescopes, particularly those with astrophotography equipment, may be able to make out surface detail on the planet. Saturn's surface is by no means as varied as Jupiter's but, like its neighbour, Saturn does have an equatorial belt. However, the lack of much colour contrast makes it trickier to pick out. There are also occasional storms in the Saturnian atmosphere that can be seen with amateur scopes.

At the observatory we field many questions about Saturn. The most frequent being, why does Saturn have rings? The honest answer is that nobody knows for sure. However, astronomers can have an educated guess. The rings are made up of small chunks of ice, each about the size of a house on average. Adding up all these pieces produces a quantity of material equivalent to one of Saturn's medium-sized moons – Mimas. So it seems conceivable that the rings may be the result of an ancient Saturnian satellite being ripped apart by the planet's gravity. If this is the case, the rings are relatively young, perhaps only a few hundred million years old.

Alternatively, a similarly sized moon could have been shattered to pieces during a feisty time in the solar system's history known as the Late Heavy Bombardment some 4 billion years ago. The same frenetic period is responsible for much of the large cratering on the Moon. However, this idea is called into question by the fact the rings are relatively bright. If they were that old, then they probably would have been dulled by the gathering of cosmic dust over time.

Although Saturn's may be the most spectacular, in fact all the gas giants – Jupiter, Saturn, Uranus and Neptune – have rings. After Saturn's, the next most extensive ring set belongs to its neighbour Uranus, which has thirteen rings. The first of them was discovered in 1977 when astronomers were watching the planet pass in front of a distant star (an event known as an 'occultation'). Uranus was able to block some of the star's light even when the main body of the planet had passed by. It was correctly assumed the planet must have rings.

International Space Station and iridium flares

We've said throughout these pages that Sirius is the brightest star in the night sky and Venus the brightest planet, and yet the brightest *object* to appear regularly (after the Moon) is actually man-made. The International Space Station (ISS), a soccer-pitch-sized satellite orbiting the Earth at 17,500mph, can shine with a maximum magnitude of -5.9. That easily trumps Venus (max -4.9) and Sirius (max -1.5).

The ISS is usually home to six people. The astronauts are treated to spectacular sights as they orbit the planet every ninety-two minutes; they can see the lights of big cities, thunderstorms dazzling like flashbulbs and the Northern and Southern Lights dancing through the atmosphere around the poles. It has been humanity's permanent dwelling in space since November 2000. Experiments conducted on board not only help us to learn about the effect of prolonged space travel on the human body (with an eye on preparing for a manned Mars mission), but also help us improve life for the rest of us left on the ground. A water filtration system designed for the ISS is now being used to provide clean water to the people of Mexico, for example. The study of how weightlessness affects the astronauts' bones is yielding insights into osteoporosis. The design of a robotic arm used on board has now been adapted to help perform brain surgery.

Due to its proximity to Earth, and its vast array of shiny solar panels, it is very easy to spot. However, its rapid speed means you'll have to be quick – it will move across your sky in a matter of minutes. It will then disappear into the Earth's shadow (knowing about this vanishing act and predicting it in front of a group can make you seem like a bit of a magician!). There are tables online showing you the exact times of ISS passes from your location (www.heavens-above.com is a great place to start).

The last few years have seen an increase in a seasonal association with the ISS. As the world above our heads has grown in popularity, so parents have been pointing out the ISS over the Christmas period and convincing their little ones that it is Santa's sleigh passing overhead.

The ISS is a wonderful sight, but some of our satellites can occasionally shine even brighter in an event known as an 'iridium flare'. At their brightest they can dazzle with a remarkable magnitude of -9.5. They are caused when sunlight glints off a satellite, causing a sudden bright light in the sky. The fleet of sixty-six Iridium communication satellites in low-Earth orbit is particularly prone to these events. As with the ISS, the orbits of iridium satellites are so predictable that we can know far in advance when a flare will be visible. Again tables of dates and times are easy to find online. However, these predictions are often specific to a certain location and you'll need to be within a few kilometres of that place in order to guarantee spotting it.

Annotated Glossary

Degree convention

Astronomers have a system for measuring the apparent size of an object in the sky, deploying the same 360-degree convention used to measure circles. A good reference point is the Full Moon, which covers half a degree. For smaller objects like stars, the degree can be further subdivided into 60 units known as arcminutes, which can in turn be split into 60 arcseconds (just like dividing up an hour of time). However, for the purposes of this book I will mostly refer to degrees.

The advantage of this system is that there are some useful ways to judge distances using your hand. First, hold up a clenched fist to the sky at arm's length (with the back of your hand facing you). The width of your fist is approximately 10 degrees. Now raise your little finger and index finger – the gap between their tips is 15 degrees. Swap your index finger for your thumb and the gap stretches to 25 degrees. You can go smaller too.

Your three middle fingers together represent 5 degrees and your little finger by itself is 1 degree wide.

Sky coordinates

Once you can use the degree system, you'll be much better placed to grapple with the celestial coordinates that astronomers use to chart the heavens. On Earth we use two coordinates – our longitude and latitude – to denote where we are on Earth's surface. Kielder, for example, is located at 55.23°N and 2.62°W. Likewise, astronomers use right ascension (RA) and declination (Dec) to describe the position of an object on a giant celestial sphere that extends outwards from the Earth. As on the Earth's surface, we mark a celestial equator on the celestial sphere by extending the Earth's equator out into space. The celestial poles match the Earth's North and South Poles too.

Let's start with RA, the equivalent of longitude on Earth. Kielder has a longitude of 2.62°W because we are located that far west of the prime meridian running through the courtyard of the Royal Observatory, Greenwich. The prime meridian is the point from which all longitude on Earth is measured. In the sky, the equivalent is the First Point of Aries. At the spring equinox in March, the Sun crosses the celestial equator at this point. Right ascension is measured from here in hours, minutes and seconds, increasing towards the east. To complicate matters slightly, the stars have drifted since the system was first devised and so the First Point of Aries is now in Pisces.

Magnitude

Magnitude refers to an object's brightness. There are two different types: absolute and apparent. An object's absolute magnitude is how bright it really is, whereas its apparent magnitude is how bright we see it in our sky.

When it comes to practical observation, it is almost always the apparent magnitude we're interested in. After all, our concern is how bright or dim it appears to our eyes, binoculars or telescopes. This magnitude system originated in Greece 2,000 years ago, with all stars visible to the unaided eye split into six magnitudes. The brightest were labelled 'first magnitude', the dimmest 'sixth magnitude'. In 1856, the system was formalised: it was stated that a sixth magnitude star is exactly 100 times fainter than a first magnitude star. Consequently, the brightness ratio between two stars in two neighbouring magnitudes is 2.5. For example, a magnitude-1.0 star is 2.5 times brighter than a magnitude-2.0 star.

The modern version of this system uses the star Vega, in the constellation of Lyra, as a reference and it is defined to have an apparent magnitude of exactly 0.0. Particularly bright objects have negative apparent magnitudes. Across both hemispheres there are only four stars in the night sky brighter than Vega: Arcturus (-0.05), Alpha Centauri (-0.27), Canopus (-0.74) and Sirius (-1.46). The Sun (-26.74), Moon (-12.74) and all five naked-eye planets also shine with negative magnitudes. Even some man-made objects do the same. Thanks to its enormous array of solar panels, the International Space Station (ISS) dazzles with a maximum magnitude of -5.9, making it brighter than any of the planets. Going the other way, higher numbers mean fainter objects. A big amateur telescope will struggle to make out objects dimmer than about magnitude 14.0, whereas the Hubble Space Telescope can push all the way up to magnitude 32.0.

Ecliptic

Of the eighty-eight constellations, people are most likely to know the names of twelve of them in particular: Aries, Taurus, Gemini, Cancer, Leo, Virgo, Libra, Scorpius, Sagittarius, Capricornus, Aquarius and Pisces. They represent the zodiac, the signs commonly associated with astrology and horoscopes. They also mark out the approximate path the Sun takes across our sky as we orbit it – a line known as the ecliptic. As the

Moon orbits the Earth, it also never strays very far from this line.

The ancients noticed another special property about the ecliptic as they watched the skies over long periods. Five stars appeared not to be fixed into constellations like the rest, and would instead wander along this line. They dubbed them 'wandering stars', or *asteres planetai* in Greek. It is from that phrase that we get our modern-day word for these wandering stars: planets. Mercury, Venus, Mars, Jupiter and Saturn are not stars at all but appear to move along the ecliptic as they also orbit the Sun. Uranus and Neptune do the same, but because of their great distance from the Sun, it required the invention of the telescope to see them. However, their wandering nature means that, unlike the constellations, the planets and the Moon aren't always visible at the same time every year. That means I can't tell you to look for Mars in March or Saturn in September. That said, many of the bi-monthly chapters contain information on some of these objects, with details of how to observe them when they are visible.

Common and scientific star names

Throughout the bimonthly night-sky guides, and the book in general, I interchange both the common and scientific names for the stars. Many of the common names are literal and derived from Arabic. The famous star Betelgeuse in Orion, for example, is often said to mean 'armpit of the Central One', although this is sometimes disputed. Scientifically they are classified using the Bayer designation. Named after German astronomer Johann Bayer, the system uses Greek letters to denote the brightness of stars in a constellation – alpha (α) for the brightest, beta (β) for the next brightest, etc. These letters are followed either by the full Latinised name of the constellation or a three-letter abbreviation. So Sirius, the brightest star in the constellation of Canis Major, is also referred to as Alpha Canis Majoris or α CMa. This is most often used by amateur astronomers when there is no common name for a star. Occasionally you will also see fainter stars named another way, such as Betelgeuse's neighbour HIP 28100. The HIP stands for Hipparcos – a catalogue of dimmer stars amassed between 1989 and 1993. This only crops up when using these stars to locate deep-sky objects such as nebulae and galaxies through telescopes.

Below is a reference list of Greek letters to help you through the sky guide sections.

α	Alpha
β	Beta
γ	Gamma
δ	Delta
ϵ	Epsilon
ζ	Zeta

η	Eta
θ	Theta
ι	Iota
κ	Kappa
λ	Lambda
μ	Mu
ν	Nu
ξ	Xi
o	Omicron
π	Pi
ρ	Rho
σ	Sigma
τ	Tau
υ	Upsilon
φ	Phi
χ	Chi
ψ	Psi
ω	Omega

A Note on Equipment

Many of those people who are beginning to get into astronomy – or indeed are being bugged by their space-mad children – automatically plump for a telescope. However, when first starting out, it can be just as worthwhile to purchase a pair of binoculars. After all, binoculars are effectively two mini telescopes placed side by side. Plus, if you're just getting to know the sky, it can be intimidating having to learn how a telescope works at the same time. It is often better to start with binoculars and then move on to a telescope once you're more confident. Having said that, please be aware that children sometimes find binoculars difficult to use.

The bi-monthly night-sky guides in this book will give you plenty of deep-sky targets to look out for with binoculars. Through a typical set you should expect to make out up to tenth magnitude objects from a dark site. That means the number of stars available to you will jump from around 3,000 with your eyes alone to more than 200,000. Casting your binoculars over the band of the Milky Way will reveal a particularly rich star field.

Binoculars aren't just about seeing more individual stars, however. The Moon, for example, is a breathtaking sight through binoculars because you can still manage to get the whole thing in the same field of vision. That's another big advantage of binoculars over telescopes – lower magnification. Many of the targets I list are even wider than the Moon. One of the most famous star clusters in the sky – the Pleiades in Taurus – spans 1.5 degrees or approximately three Full Moon widths. Even through a low-power telescope you would struggle to get all the stars in the same field of view. Seeing only a bit of it at a time isn't quite as spectacular. A pair of 10 x 50 binoculars on the other hand, typically with a 6- or 7-degree field of view, would be ideal. The Andromeda galaxy, another popular target for beginners, has a diameter of 3 degrees and so is also better seen through binoculars, or a small telescope.

The pages in this book describe in more detail the nuances of different telescopes, and what is suitable for both a beginner and someone looking to progress.

List of Constellations

Constellation	Meaning	Three-Letter Abbreviation	Latinised Form
Andromeda	Chained lady	And	Andromedae
Antilia	Air Pump	Ant	Antliae
Apus	Bird of Paradise	Aps	Apodis
Aquarius	Water-bearer	Aqr	Aquarii
Aquila	Eagle	Aql	Aquilae
Ara	Altar	Ara	Arae
Aries	Ram	Ari	Arietis
Auriga	Charioteer	Aur	Aurigae
Boötes	Herdsman	Boo	Boötis
Caelum	Engraving Tool	Cae	Caeli
Camelopardalis	Giraffe	Cam	Camelopardalis
Cancer	Crab	Cnc	Cancri
Canes Venatici	Hunting Dogs	CVn	Canum Venaticorum
Canis Major	Larger Dog	CMa	Canis Majoris
Canis Minor	Smaller Dog	CMi	Canis Minoris
Capricornus	Water Goat	Cap	Capricorni

(*Continued*)

Constellation	Meaning	Three-Letter Abbreviation	Latinised Form
Carina	Keel	Car	Carinae
Cassiopeia	Queen	Cas	Cassiopeiae
Centaurus	Centaur	Cen	Centauri
Cepheus	King	Cep	Cephei
Cetus	Whale	Cet	Ceti
Chameleon	Chameleon	Cha	Chamaeleontis
Circinus	Compasses	Cir	Circini
Columba	Dove	Col	Columbae
Coma Berenices	Berenice's Hair	Com	Comae Berenices
Corona Australis	Southern Crown	CrA	Coronae Australis
Corona Borealis	Northern Crown	CrB	Coronae Borealis
Corvus	Crow	Crv	Corvi
Crater	Cup	Crt	Crateris
Crux	Southern Cross	Cru	Crucis
Cygnus	Swan	Cyg	Cygni
Delphinus	Dolphin	Del	Delphini
Dorado	Swordfish	Dor	Doradus
Draco	Dragon	Dra	Draconis
Equuleus	Little Horse	Eql	Equulei
Eridanus	River	Eri	Eridani
Fornax	Furnace	For	Fornacis
Gemini	Twins	Gem	Geminorum
Grus	Crane	Gru	Gruis
Hercules	Hero	Her	Herculis
Horologium	Clock	Hor	Horologii
Hydra	Water Serpent	Hya	Hydrae
Hydrus	Water Snake	Hyi	Hydri

Constellation	Meaning	Three-Letter Abbreviation	Latinised Form
Indus	Indian	Ind	Indi
Lacerta	Lizard	Lac	Lacertae
Leo	Lion	Leo	Leonis
Leo Minor	Smaller Lion	LMi	Leonis Minoris
Lepus	Hare	Lep	Leporis
Libra	Scales	Lib	Librae
Lupus	Wolf	Lup	Lupi
Lynx	Lynx	Lyn	Lyncis
Lyra	Lyre	Lyr	Lyrae
Mensa	Table	Men	Mensae
Microscopium	Microscope	Mic	Microscopii
Mononceros	Unicorn	Mon	Monocerotis
Musca	Fly	Mus	Muscae
Norma	Square	Nor	Normae
Octans	Octant	Oct	Octantis
Ophiuchus	Serpent-bearer	Oph	Ophiuchi
Orion	Hunter	Ori	Orionis
Pavo	Peacock	Pav	Pavonis
Pegasus	Winged Horse	Peg	Pegasi
Perseus	Hero	Per	Persei
Phoenix	Phoenix	Phe	Phoenicis
Pictor	Easel	Pic	Pictoris
Pisces	Fishes	Psc	Piscium
Piscis Austrinus	Southern Fish	PsA	Piscis Austrini
Puppis	The Poop Deck	Pup	Puppis
Pyxis	Compass	Pyx	Pyxidis
Reticulum	Net	Ret	Reticuli

(*Continued*)

Constellation	Meaning	Three-Letter Abbreviation	Latinised Form
Sagitta	Arrow	Sge	Sagittae
Sagittarius	Archer	Sgr	Sagittarii
Scorpius	Scorpion	Sco	Scorpii
Sculptor	Sculptor's Studio	Scl	Sculptoris
Scutum	Shield	Sct	Scuti
Serpens	Serpent	Ser	Serpentis
Sextans	Sextant	Sex	Sextantis
Taurus	Bull	Tau	Tauri
Telescopium	Telescope	Tel	Telescopii
Triangulum	Triangle	Tri	Trianguli
Triangulum Australe	Southern Triangle	TrA	Trianguli Australis
Tucana	Toucan	Tuc	Tucanae
Ursa Major	Greater Bear	UMa	Ursae Majoris
Ursa Minor	Smaller Bear	UMi	Ursae Minoris
Vela	Sails	Vel	Velorum
Virgo	Maiden	Vir	Virginis
Volans	Flying Fish	Vol	Volantis
Vulpecula	Fox	Vul	Vulpeculae

Where Can I Get More Help?

When it comes to getting to know the night sky, nothing beats practice and familiarity. Luckily there are many tools out there to help you.

If you want an easy-to-use guide to the entire sky, you could pick up a planisphere, a cheap star map made from two circular card discs on top of one another. You'll notice the months around the rim – you need to twist it round to the appropriate time of year. The constellations visible at that time then appear in the window. It's worth saying that each planisphere is designed for a particular latitude, so taking a London planisphere (51.5°N) down to Sevilla in southern Spain (37.4°N) wouldn't show you all the right constellations.

You could also pick up a popular astronomy magazine. In the UK you can choose from *Astronomy Now*, *BBC Sky at Night* or *All About Space* (all monthly). In the US, titles include *Sky & Telescope* and *Astronomy*. Most contain a guide to that month's night sky, along with practical observing tips and reviews of binoculars, telescopes and other observing equipment.

Planetarium software is also an increasingly popular resource. Stellarium (www.stellarium.org), for example, is a fantastic (and free) program which shows you the night sky for any past, present or future date for any location on Earth. The easy-to-use interface allows you to mark the constellations and search for particular stars and deep-sky objects. Clicking on any object reveals a wealth of information including its magnitude, distance and RA/Dec and Alt/Az coordinates.

In a similar vein, there are many free or low-cost apps you can download for smartphones and tablets. Some even use the device's GPS tracker to overlay a depiction of the constellations as you point it towards the real night sky. This can be particularly useful when combined with traditional sky maps, as you can first try to find the constellation for yourself, and then use the app to check you're looking in the right place.

Acknowledgements

Writing this book has been one of the most challenging and formative experiences of my life.

Thanks to Jamie Doward from the *Observer*, for writing an article last year about Kielder Observatory that was read by Ben Brusey, an editor at Century who subsequently chased me up and asked me to write my own story. Without Jamie's article, this book wouldn't have happened.

Special thanks to Ben who has guided me through the process of writing the book and has been a pillar of support, as well as a source of creativity. Without doubt, the whole team at Century has been fantastic, offering me so much professional help and advice. Many thanks also to Colin Stuart for his help with the science sections, and to Darren Bennett for drawing the star charts.

To Peter Sharpe and the team for raising the funds to build Kielder Observatory in the first place, and for having the foresight to see how the facility could benefit the community.

To Lynn and Kevin Baxter, Graham Darke, Malcolm Robinson, Austin Bowman and Paul Lewis, without whom Kielder Observatory would not have achieved its success. To Jurgen Schmoll, a friend and confidant, as well as a kick-ass instrument scientist who has helped me develop the facility. To all of the guys and girls at SAS, who I regard as astronomical evangelists. To Don Smith and David Sinden, who aren't with us anymore.

To Professor Sir Arnold Wolfendale and all of the extraordinary staff at Durham University who have welcomed me into their 'family'; you have been a constant source of help and encouragement for me to study and learn.

Special thanks to my wonderful team at Kielder, who help deliver our message to guests almost every night of the year. Thanks to every single man, woman and child who has volunteered. Thank you, too, to the trustees of KOAS, who give up their time and expertise to plan the observatory's future.

Finally, to my family: Mam, Dad and my amazing three sons, without whom my life would be meaningless. To my beautiful daughter for making grey skies brighter. To Maureen, for her tireless dedication in supporting me and our family for so long. To Lyra my new dog; to my partner Sarah who helped me set up the running of the observatory in the dim past, and is a constant source of inspiration in how to live my life. Your support Sarah is as timeless and never ending as the universe itself.

Thank you, all. You have helped make my dreams come true.